D1349061

a30213 003147157b

STATISTICAL MECHANICS AT THE TURN OF THE DECADE

Dr. George E. Uhlenbeck
Professor, The Rockefeller University

STATISTICAL MECHANICS AT THE TURN OF THE DECADE

Edited by

E. G. D. COHEN

THE ROCKEFELLER UNIVERSITY
NEW YORK, NEW YORK

MARCEL DEKKER, INC., New York 1971

MARCEL DEKKER, INC., 95 Madison Avenue, New York, N.Y. 10016

LIBRARY OF CONGRESS CATALOG CARD NUMBER 70-134789

ISBN 0-8247-1111-4

PRINTED IN THE UNITED STATES OF AMERICA

PREFACE

To commemorate the seventieth birthday of Professor G. E. Uhlenbeck, Professor A. J. F. Siegert organized on October 28 and 29, 1969 a Symposium "Statistical Mechanics at the Turn of the Decade" at Northwestern University, Evanston, Illinois. At this Symposium eight invited lectures, of a duration of fifty minutes each, were given, each followed by a discussion.

Since the lectures seemed worthwhile to preserve, I took it upon myself to have them published. They are printed here essentially as they were given at the Symposium. No attempt was made to rework them in the form of complete survey articles; therefore, the style was kept as in the original presentations, giving them a similar freshness as when delivered. Some of the articles contain material that cannot or can only with difficulty be found in the literature; others have been published in a similar form before. I felt, however, that *all* lectures should be included, regardless of the novelty of the material presented, in order to depict the Symposium as a whole. In this connection I am particularly indebted to Professors P. C. Martin and P. C. Hohenberg for their willingness to have their contributions published in this volume, in spite of the fact that similar ones had appeared previously. Although the eight lectures published here clearly do not survey all of statistical mechanics, they at least cover an important part of it and reflect, perhaps, the current status of the field.

It is hard to find a person who, over a period of almost 45 years, has been more inspiring and stimulating in the field of statistical mechanics than Professor Uhlenbeck. Through his continual insistence on clarity and deeper understanding, he has ennobled the field, and it is difficult to decide whether his most significant contribution to statistical mechanics is to be found in his fundamental papers, or in the level of his criticism, which gives him—as it does his teacher Professor P. Ehrenfest—a unique position in statistical mechanics.

It is to him that this book is dedicated in gratitude and admiration.

E. G. D. Cohen

CONTRIBUTORS TO THIS VOLUME

E. G. D. Cohen–The Rockefeller University, New York, New York

C. Domb–King's College, University of London, London, England

Freeman J. Dyson–Institute for Advanced Study, Princeton, New Jersey

P. C. Hohenberg–Bell Telephone Laboratories Inc.,
Murray Hill, New Jersey

Paul C. Martin–Harvard University, Cambridge, Massachusetts

D. Ruelle–Institut des Hautes Etudes Scientifiques,
Bures sur Yvette, France

A. J. F. Siegert–Northwestern University, Evanston, Illinois

A. S. Wightman–Princeton University, Princeton, New Jersey

CONTENTS

STATISTICAL MECHANICS AND ERGODIC THEORY: AN EXPOSITORY LECTURE

A. S. Wightman

Princeton University
Princeton, New Jersey

I want to make it clear at the outset that this lecture contains absolutely nothing new for the experts; it is an attempt to explain in an elementary way the significance for <u>classical</u> statistical mechanics of some mathematical developments of the last decade. Perhaps it is not entirely inappropriate that such an expository lecture appear in a symposium in honor of Professor Uhlenbeck. To the delight and enlightenment of several generations of students of statistical mechanics he has regarded it as his responsibility to make the subject as clear as he could.

I. DYNAMICAL SYSTEMS AS MODELS FOR STATISTICAL MECHANICS

Maxwell, Boltzmann, Gibbs and others of the founders

of statistical mechanics posed fundamental problems[1]:

How can microscopically highly complex motions of atoms give rise to macroscopic phenomena with simple deterministic descriptions?

How can microscopic reversibility give rise to apparent macroscopic irreversibility?

Their profound answers involve a complicated mixture of arguments which are based, at least in part, on the following factors:

a) The orbits of dynamical systems in phase space are highly sensitive to initial conditions. They wander everywhere that the conservation laws for macroscopic quantities permit them to. (Ergodic Hypothesis)

b) The number of particles N is so large that it is only the limit as $N \to \infty$ that counts in the answer to physical questions. (Thermodynamic Limit)

c) The observable quantities relevant for statistical mechanics are macroscopic observables.

d) The probability of departure from average behavior is very small, being typically proportional to some inverse power of the number of particles in the system.

Even this fragmentary list makes it clear that the study

of the flow in a phase space defined by an isolated Hamiltonian system is not going to give the whole answer to the fundamental problems of classical statistical mechanics. There are those who would go much further and argue that the results of such a study are necessarily irrelevant for the fundamental problems.[2]

One can make a three-fold response to this attitude. First, the founders of statistical mechanics were not so prudish; they regarded the theory of isolated dynamical systems to be of interest. Second, if one does not understand what actually happens in such systems, one cannot be sure that the introduction of macroscopic observables or the passage to the thermodynamic limit is essential for the appearance of macroscopic determinism and irreversibility. Third, many actual systems studied in the laboratory are approximately isolated Hamiltonian systems. In any case, the attitude taken here is that it is worthwhile finding out what goes on in isolated Hamiltonian dynamical systems. In fact, such systems sometimes _do_ exhibit striking features of "irreversibility" that make them relevant for statistical mechanics, as I will soon describe.

II. STUDIES OF THE QUALITATIVE PROPERTIES OF HAMILTONIAN FLOWS IN ANCIENT TIMES

The flow in phase space associated with a Hamiltonian system is the family of curves of $\{q(t), p(t); -\infty < t < \infty\}$ for all possible initial conditions. (For some purposes it is also natural to restrict attention to a single energy surface $H(q, p) = \text{const}$, and we will do so on occasion.)

It was Poincaré who was the founder and pioneering figure of the qualitative theory of dynamical systems. He asked and developed methods for answering such questions as: How many periodic orbits are there? Does almost every point on an energy surface return arbitrarily close to itself (a recurrence property)? If one knows a periodic orbit what is the qualitative behavior of the orbits that are nearby at some particular time?[3]

Let me describe how Poincaré went about analyzing the last problem. He introduced the notion of a transversal section of a flow; that is, a surface nowhere tangent to the flow, as illustrated in Fig. 1.[4]

Associated with a transversal section there is a mapping of the transversal section into itself obtained by letting $x \to \phi(x)$ where $\phi(x)$ is the point x arrives at the first

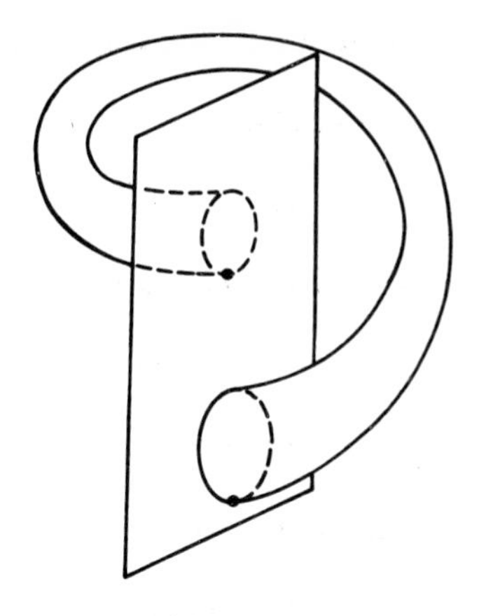

FIG. 1

A transversal section.

time it passes through the transversal section again. Clearly, if a point x_o of the transversal section lies on a periodic orbit, it is a fixed point of ϕ (or possibly some iterate of ϕ ; for simplicity, we will ignore this latter possibility):

$$\phi(x_o) = x_o$$

and by continuity all points x of the transversal section sufficiently close to x_o will be mapped again into the transversal section. Such a mapping ϕ is called the Poincaré map associated with a periodic orbit and transversal section.

Consider now the action of ϕ on points in the neighborhood of a fixed point x_o. In order to get a preliminary classification of ϕ's behavior, we approximate ϕ by the linear terms in its Taylor series evaluated at x_o, i.e., ϕ is replaced by the linear transformation $\nabla \phi(x_o)$. (This is equivalent to expanding the Hamiltonian in a power series around the orbit in question and keeping only quadratic terms.) The natural objects to investigate are the eigenvalues of $\nabla \phi(x_o)$. They turn out to be independent of the choice of the coordinates in the neighborhood of x_o and of the transversal section. They are called characteristic multipliers for the periodic orbit.

Clearly, if an eigenvalue λ of $\nabla \phi(x_o)$ has absolute value $|\lambda| > 1$, it is an indication of instability since the iterates of $\nabla \phi(x_o)$ will then have arbitrarily large eigenvalues. Conversely, eigenvalues such that $|\lambda| < 1$ should be associated with orbits that are asymptotic to x_o. For $|\lambda| = 1$, one can expect bounded oscillations around x_o. These indications from the linear approximation to ϕ can be extended to a statement about the behavior of the exact orbits under ϕ. There is a manifold of orbits called the unstable manifold generated by the vectors belonging to the eigen subspaces whose eigenvalues have $|\lambda| > 1$. There is a stable manifold of orbits related analogously to the eigenvalues with $|\lambda| < 1$, and a center manifold associated with $|\lambda| = 1$. Intuitively in a two-dimensional picture various cases are indicated in Fig. 2.[5]

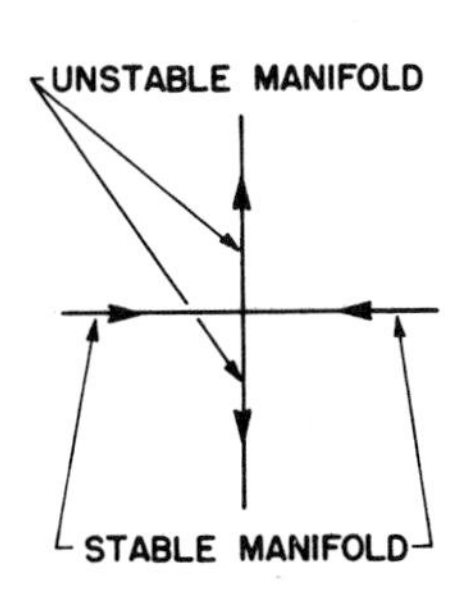

CASE A: $|\lambda_1| < 1$ $|\lambda_2| > 1$ CASE B: $|\lambda_1| = 1$ $|\lambda_2| = 1$

FIG. 2

Behavior of orbits in the neighborhood of a periodic orbit. The pictures show a transversal section of an energy surface for a system of two degrees of freedom.

One might think that generic behavior for a dynamical system should correspond to all $|\lambda| \neq 1$, and that is indeed the case if one admits non-Hamiltonian systems. But it is a striking fact that for Hamiltonian systems the eigenvalue 1 has odd multiplicity (and is therefore of multiplicity ≥ 1). This is a simple consequence of the fact that for a Hamiltonian system $\nabla \phi (x_0)$ is the restriction of an infinitesimal symplectic transformation to the transverse section. In this sense Hamiltonian systems are very special. Liapounov already exploited this fact to show the existence of a one-parameter family of periodic orbits in some neighborhood of a given periodic orbit of a Hamiltonian system.[6,7] Related elementary facts about the eigenvalues of symplectic transformations show that the dimensions of the unstable manifold and stable manifold are the same. Thus this analysis shows that, in a Hamiltonian system, for all orbits in some neighborhood of the given periodic orbit to be stable one must have $|\lambda| = 1$ for all characteristic multipliers λ. This we will refer to as the _pure center case_.

The study of the behavior of the orbits in the neighborhood of a periodic orbit is a famous problem of such a magnitude as to engage much of the attention of Poincaré in the last years of his life. (The celebrated last theorem of Poincaré states that there exist an infinite number of periodic orbits in every neighborhood of such a periodic orbit for a Hamiltonian system of two degrees of freedom. Poincaré left one essential step in his proof unfinished. The gap was filled by a theorem of G. D. Birkhoff.[8] However, neither the work of Poincaré nor the later work of Birkhoff settled the stability of the majority of orbits near the given periodic orbit. It is just this subject in which there have been remarkable new developments in the last decade and a half, about which I want to speak later.

Before turning to these more recent results, I want to complete the setting by describing old work on some other aspects of the theory of dynamical systems. The early work in statistical mechanics left the mathematical status of the ergodic hypothesis quite unclear. In the early thirties, the papers of von Neumann, Birkhoff, Khintchin, E. Hopf, and others created a new chapter of analysis, ergodic theory, and opened the way for a thorough mathematical elaboration of the underlying ideas.[9]

It turned out that what was essential for the first theorems was that the flow under consideration possess an <u>invariant measure</u>, μ. (In the Hamiltonian case, μ is just the volume in phase or, for the restriction to an energy surface, the corresponding surface area.) Then in the modern terminology the flow is called <u>ergodic</u> if the space cannot be partitioned into two sets invariant under the flow and of measure greater than zero.[10] The main theorems of von Neumann and Birkhoff asserted two things. First, that the time average

$$\lim_{T\to\infty} \frac{1}{2T} \int_{-T}^{T} f(x_t)dt = \bar{f}(x)$$

exists. (In von Neumann's case f is assumed to be square integrable

$$\int |f(x)|^2 \, d\mu(x) < \infty ,$$

and the convergence to the limit is in the mean:

$$\lim_{T\to\infty} \int d\mu(x) \left| \frac{1}{2T} \int_{-T}^{T} f(x_t) \, dt - \bar{f}(x) \right|^2 = 0 ;$$

while in Birkhoff's case f is assumed to be integrable

$$\int |f(x)| \, d\mu(x) < \infty ,$$

and the convergence is pointwise almost everywhere.) Second, if f is integrable and $\int d\mu(x) < \infty$, then

$$\int \bar{f}(x) \, d\mu(x) = \int f(x) \, d\mu(x) ,$$

and if the flow is ergodic $\bar{f}$ is constant almost everywhere and

$$\bar{f}(x) = [\int d\mu(x)]^{-1} \int f(x) \, d\mu(x) ;$$

i.e., time averages are equal to space averages.

In passing, it is worth noting that these theorems have discrete analogues in which one considers the successive iterates of a single transformation T and replaces the time average (1) by

$$\lim_{N\to\infty} \frac{1}{2N} \sum_{j=-N}^{N} f(T^j x) .$$

Shortly after von Neumann and Birkhoff obtained their results an important refinement of the notion of ergodic flow was introduced: a mixing flow.[11] For simplicity consider a flow for which the total space X has finite measure. Then the flow is mixing if for any two measurable sets A and B

$$\lim_{t\to\infty} \mu(A \cap B_t) = \frac{\mu(A)\,\mu(B)}{\mu(X)} .$$

It turns out that every mixing flow is ergodic. Hopf pointed out that ergodicity of a flow does not suffice in order that a flow should simulate an honest roulette wheel but the mixing property does. For reasons that escape me completely Hopf's beautiful and simple results on mixing and its physical significance do not seem to have been assimilated into the pedagogical literature of theoretical physics at all. Because I want to leave you with an indelible intuitive feeling for the distinction between ergodicity and mixing, let me describe two very simple flows the first of which is ergodic but not mixing, while the second is mixing. (More accurately, what I shall describe is the mapping that is the result of letting the flow continue for one second). The phase space is a two-dimensional torus or what is the same thing the unit square with periodic boundary conditions. The first mapping is

$$(x, y) \rightarrow (x+1, \ y+\alpha) \quad \text{mod } 1 \ ; \tag{2}$$

the second is

$$(x, y) \rightarrow (x+y, \ x+2y) \quad \text{mod } 1 \ . \tag{3}$$

The first merely translates the cat along a direction whose slope is α and does not change its shape. (See Fig. 3.) If α is irrational the flow is ergodic but not mixing.

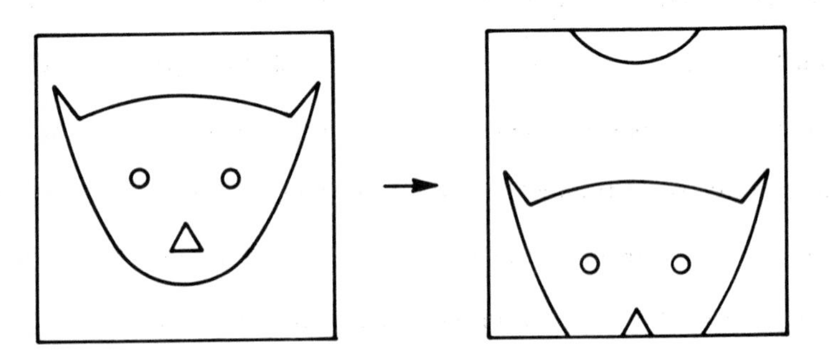

FIG. 3

The transformation (2).

For ergodicity it suffices if cats spend equal amounts of time in equal volumes of phase. On the other hand, the second transformation does much more radical things to the cat. (See Fig. 4.) You will have to admit that after the second iteration of the transformation we already have a rather mixed up cat.

Notice that every orbit of the flow associated with (2) is dense, and because it is ergodic, time averages of functions defined on the phase space are equal to phase space averages.

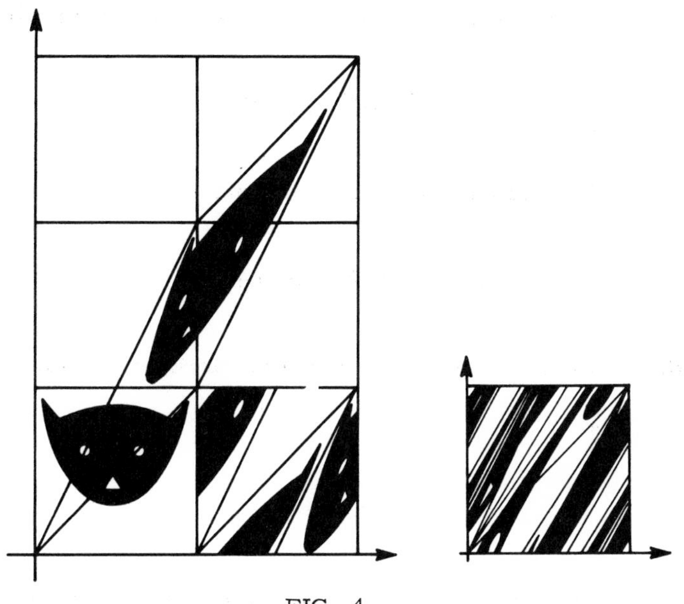

FIG. 4

The transformation (3) and its square.

Nevertheless, the flow does not have the kind of "irreversibility" ($\equiv$ instability) that one associates with physical systems like a stirred pitcher of initially unmixed gin and vermouth. If two points of the phase space are close together they stay close together for all time in the flow associated with (2).

There is an auxiliary formalism proposed by Koopman, which gives a very succinct expression for the distinction between ergodicity and mixing. Koopman noted that the functions f defined on the space X where the flow acts, that are square integrable with respect to the invariant measure μ

$$\int |f(x)|^2 \, d\mu(x) < \infty$$

form a Hilbert space with scalar product

$$(f, g) = \int \overline{f(x)} \, g(x) \, d\mu(x) \quad .$$

He then considered the one-parameter family of operators V^t defined by

$$(V^t f)(x) = f(x_t) \ .$$

They are unitary because the flow is measure preserving, and form a group

$$V^{t_1} V^{t_2} = V^{t_1 + t_2}$$

because $(x_{t_1})_{t_2} = x_{t_1 + t_2}$. Under very general smoothness assumptions on the dependence of x_t on t and x, V^t turns out to be continuous in t. Thus, one can apply the theory of continuous one-parameter unitary groups (Stone's theorem) to obtain

$$V^t = \exp\ iAt \quad ,$$

where A is some self-adjoint operator. The idea is then to characterize the flow in terms of properties of the operator A.

Consider the case $\mu(X) < \infty$. Then the constant function is always an eigenfunction of A of eigenvalue 0 because V^t leaves such functions invariant. The flow is ergodic if and only if constant functions are the only invariant functions, i.e., if and only if 0 is a simple eigenvalue of A.

Can A have eigenvalues other than 0, if the flow is ergodic? The answer is yes in general, but no if the flow is mixing. When the flow is mixing A has one simple eigenvalue, 0, and all the rest of its spectrum is continuous. The converse is not quite true. If A has 0 as a simple eigenvalue and no other eigenvalues, one gets what is called the <u>weak mixing</u> property

$$\lim_{n\to\infty} \frac{1}{n} \sum_{j=0}^{n-1} \left| \mu(A_{jt} \cap B) - \frac{\mu(A)\mu(B)}{\mu(X)} \right| = 0 .$$

On the other hand, if one is willing to assume more about the continuous spectrum of A; namely, that A has a homogeneous Lebesgue spectrum one can prove the mixing property. Homogeneous Lebesgue spectrum means that when one diagonalizes A every real number λ lies in the spectrum and has the same multiplicity and that the spectral weight is just $d\lambda$.

We see here the first inklings of a hierarchy of properties that characterize a flow as being more and more "irreversible":

ergodic $\leftarrow$ weak mixing $\leftarrow$ mixing $\leftarrow$ homogeneous Lebesgue spectrum

The last family of ideas of the thirties that I want to mention is a response to the question: what local properties of a flow guarantee that it is ergodic or mixing. Work in this direction by Hopf, Hedlund, and others began with a study of geodesic flow on surfaces of constant negative curvature. The reason for singling out these dynamical systems for a careful study was simple. From studies by Hadamard,[12] Artin,[13] and others,[14] it was known that the orbits are wildly unstable, so the systems provide very good candidates for ergodic or mixing flows.

A first elementary point: do not be confused by the fancy terminology that refers to the geodesic flow in the unit tangent bundle of an n-dimensional Riemannian manifold V. That is just the frictionless motion of a particle on V where, because the kinetic energy is a constant of the motion, the velocity has been normalized to 1. If one does not normalize, one is dealing with a Hamiltonian system of n degrees of freedom. Restriction to the energy surface $|v|=1$ ($\equiv$ unit tangent bundle) yields a flow on a manifold of $2n-1$ dimensions.

In the simple case $n=2$ there are two geometric features of the flow which Hopf found to provide a key to the ergodic and mixing properties of the flow: asymptotes and horospheres. The ideas involved are as follows. Suppose $\gamma = \{q(t); -\infty < t < \infty\}$ is a geodesic in the manifold V, and q' is any point of V. Then by virtue of the special properties of a manifold of negative curvature, there is a unique geodesic through q' and $q(T)$, call it $\gamma(q', T) = \{q'(T, t); -\infty < t < \infty\}$. Now let T approach $+\infty$. Then one proves that there is a unique limiting geodesic

$\gamma(q', \infty) = \{q'(\infty, t); -\infty < t < \infty\}$. It is called the + _asymptote to_ γ _through_ q' . There is an analogous construction of the - _asymptote_ as $T \to -\infty$. Now consider all the + asymptotes of the geodesic γ . The surfaces normal to all these geodesics are called + _horospheres_ of γ . Similarly, one can construct - _horospheres_ (See Fig. 5).

It turns out that the flow carries horospheres into horospheres and that points $\{q, p\}$ of phase space lying on the same + horosphere approach each other with exponential rapidity as $t \to +\infty$ and diverge from each other with exponential rapidity as $t \to -\infty$. It gives quantitative expression to the instability of the flow. Hopf showed that it led to elegant proofs of the ergodicity and mixing properties for two- and then n-dimensional V and constant negative curvature. He then discussed the cases of variable curvature and isolated an important property which I shall discuss later.

Amazingly enough, although these ideas have great intuitive appeal, only one physicist tried to follow them up to see whether systems of more direct physical interest (say hard spheres in a box) are mixing. That was N. S. Krylov[15] who unfortunately died young (1947) and whose work only became generally known as a result of the work of J. Sinai

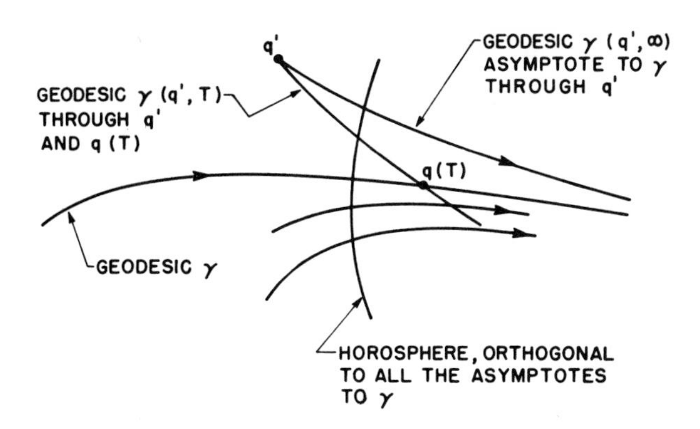

FIG. 5

Asymptotes and horospheres.

many years later. Krylov's idea was that for hard spheres in a box a very small change in direction for a particle suffering a collision results in quite large uncertainties in direction after a few collisions. He gave heuristic but interesting arguments that as a result the flow defined by hard spheres in a box is mixing.

Enough for ancient times.

III. STABILITY OF ORBITS IN THE PURE CENTER CASE

In the above review I mentioned that the stability of orbits in the neighborhood of a periodic orbit for a pure center (all characteristic multipliers of absolute value 1) was left open by the grand masters of the subject, Poincaré, Birkhoff et al.

It was taken up again by A. Kolmogorov whose address to the Congress of Mathematicians in Amsterdam in 1954 started a new chapter in the subject.[16] Important new results significant for statistical mechanics were obtained by Arnold,[17] Moser,[18] and others.

The conclusions can be summarized as follows. In a sufficiently small neighborhood of a periodic orbit all of whose characteristic multipliers are of absolute value 1 (pure center case), and satisfying a certain algebraic nondegeneracy condition, there is an infinite family of invariant tori of dimension n if the phase space has dimension 2n. The measure of the union of these tori is greater than zero. In fact, the ratio of that measure to the total measure of the neighborhood approaches 1 as the neighborhood gets smaller. On the other hand, the orbits that are not invariant tori form a dense set, which is also of positive measure. Consequently, Hamiltonian dynamical systems possessing such periodic orbits are never ergodic. The lack of ergodicity is particularly clear for two degrees of freedom, where the invariant tori themselves separate the energy surface into invariant measurable subsets.

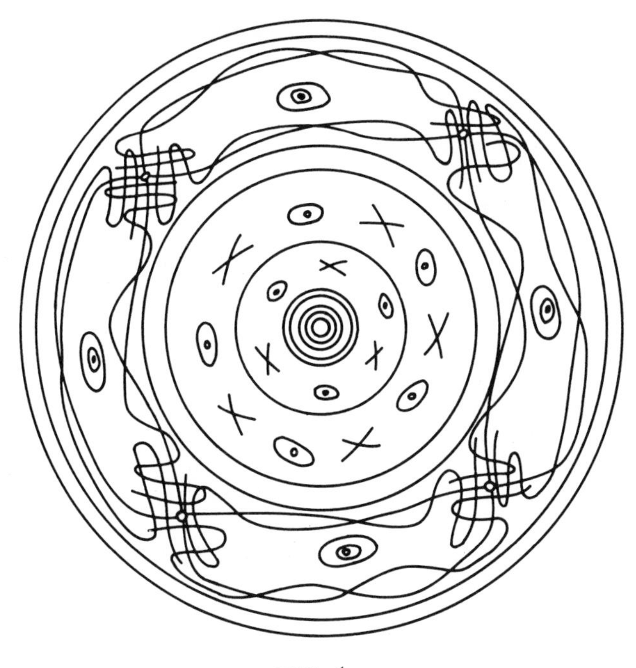

FIG. 6

Transverse section near an orbit in the pure center case.

The situation is indicated in Fig. 6, for $n = 2$. The given periodic orbit is indicated by the point in the center. It is surrounded by invariant tori. For $n = 2$ one knows that there are infinitely many other periodic orbits between the invariant tori belonging to the first periodic orbit. However, it is not known what fraction of them are in the pure center case. If, as seems possible, there is an infinitity of such orbits, then each is surrounded by its invariant tori and so on world within world. I must say that I find the fact that two oscillators with polynomial coupling can produce such phenomena simply fantastic.

There are many things that one still would like to know about the structure of the flow. For example, how does the flow on the complement of the tori decompose into ergodic components? Further, for $n > 2$, are the orbits not on these invariant tori stable? (For $n > 2$, the existence of the invariant tori does not prevent other orbits from wandering away because the tori only have dimension n.)[19]

It is not yet known whether the existence of these phenomena could have some relevance for the classical statistical mechanics of gases, etc. I will say something more about that in connection with the work of Sinai. But there is no doubt that they are relevant in celestial mechanics.

In fact, some of the interesting indications for future progress in the subject come from numerical computations whose results appear in the Astronomical Journal. Let me show one such, Fig. 7 taken from the work of Hénon and Heiles.[20]

Interestingly enough, one can get approximate formulas for the invariant tori by the same methods that yield the

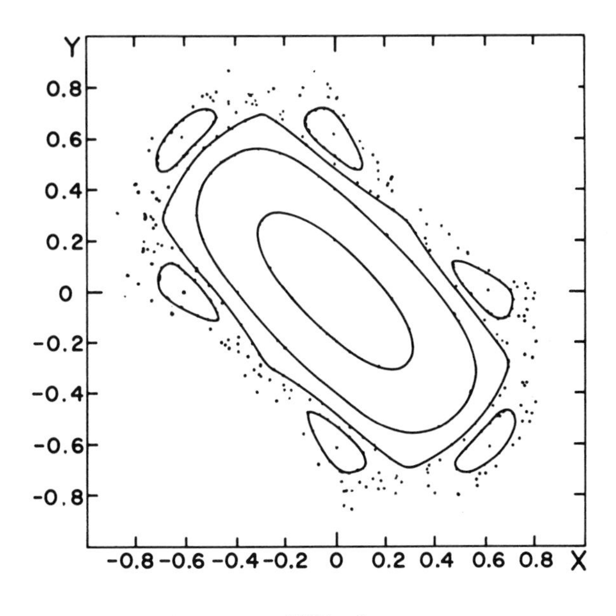

FIG. 7

Points lying on the same curves belong to the same orbit. The scattered points all belong to the same orbit.

existence proofs. (The proofs depend on making a sequence of canonical transformations which almost bring the system into a quasi-periodic system and then estimating remainders. If one throws away the remainders, one gets the required approximate formulas.)

IV. THE K-S ENTROPY; K-SYSTEMS; C-SYSTEMS

Kolmogorov also played an important role in the next development I want to describe. In his lectures in the middle 1950's, he proposed the application of the ideas of information theory to dynamical systems and made the first significant steps in that direction. His proposal was soon taken up by Sinai and others, and their results provide a whole new family of quantitative and qualitative results in the theory of dynamical systems.

The basis idea is simple. According to information theory[21] if one is given a set of n alternatives of probability $p_1 \dots p_n$, respectively, one attributes an information $-\log p_j$ to the observation of the jth alternative in an experiment. The average information associated with an experiment is thus

$$H = -\sum_{j=1}^{n} p_j \ln p_j \ .$$

If we apply this definition to a partition of phase space $\mathcal{a} = A_1, \dots, A_n$ into disjoint sets $A_i \cap A_j = \phi$, $i \neq j$ and equate p_j to the measure $\mu(A_j)$ we have associated an entropy $H(\mathcal{a})$

$$H(\mathcal{a}) = -\sum_{j=1}^{n} \mu(A_j) \ln \mu(A_j)$$

with a partition $\mathcal{a}$.

Now given two partitions $\mathcal{a}$ and $\mathcal{B}$, there is a new partition $\mathcal{a} \vee \mathcal{B}$ whose sets are

$$\{A_i \cap B_j \, ; \, A_i \in \mathcal{a} \, , \quad B_i \in \mathcal{B} \, \}$$

with the rule that a set is to be omitted if it is empty or of zero measure.

If T_t defines the action of the flow through a time t, then the sets of the partition

$$\mathcal{A} \vee T_t \mathcal{A} \vee T_{2t} \mathcal{A} \vee \cdots \vee T_{nt} \mathcal{A}$$

display how successful the flow is in mixing up the original partition $\mathcal{A}$. Kolmogorov's idea was to obtain a quantitative measure of this mixing as the average information per step over the whole flow

$$H(\mathcal{A}, T_t) = \lim_{n \to \infty} \frac{1}{n} H(\mathcal{A} \vee T_t \mathcal{A} \vee T_{2t} \mathcal{A} \vee \cdots \vee T_{nt} \mathcal{A}).$$

If the flow is very mixing, this number should be strictly positive. If that is true for every nontrivial partition $\mathcal{A}$, then the system is called a <u>K-system</u> and the flow, a <u>K-flow</u>. One can go further and define

$$h(T_t) = \sup_{\mathcal{A}} H(\mathcal{A}, T_t),$$

$h(T_t)$ is called the <u>K-S entropy</u>[22] of the flow. (S is for Sinai who simplified Kolmogorov's original definition and developed the theory extensively.) It turns out that $H(T_t)$ is proportional to t:

$$H(T_t) = ht$$

so $\frac{1}{h}$ is a <u>relaxation time</u>, the longest relaxation time associated with all the possible properties of the system.

It also turns out that every K-system is ergodic and mixing. It is not unreasonable to think that K-systems possess the typical kind of irreversibility which make them suitable subjects for statistical mechanics. The theory of the K-S entropy ought to be regarded as a first step (a very small step because the K-S entropy answers much too general a question) in the direction of a rigorous theory of the approach to equilibrium.

The work on the general theory of K-systems is characteristically measure theoretical. One can also ask which smooth dynamical systems are K-systems. Here important work of Anosov has isolated a large class of such systems: the so-called <u>C-systems</u> (that is the terminology of Arnold and Avez;[3] in Smale[3] they are called Anosov flows). A C-system is a smooth flow on a manifold such that the tangent space to the manifold at a point can be split into three subspaces, one along the flow of dimension one, and two others, the <u>contracting</u> and <u>dilating</u> subspaces. The latter are characterized by the fact that the tangent vector v_t satisfies

$$ae^{-ct}\|v_o\| \leq \|v_t\| \text{ for } t<0 \text{ and } \|v_t\| \leq be^{-ct}\|v_o\| \text{ for } t>0$$

for the contracting subspace and

$$a_1 e^{+ct}\|v_o\| \leq \|v_t\| \text{ for } t>0 \text{ and } \|v_t\| \leq b_1 e^{+ct}\|v_o\| \text{ for } t<0,$$

for the dilating subspace. Here v_t is the tangent vector at time t that arises from v_o under the action of the flow.

The dilating and contracting subspaces are a generalization of the horospheres studied by Hopf in the special case of manifolds of negative curvature and discussed above.

Sinai has proved, under a restriction that I won't go into here, that every C-system is a K-system.[23]

Thus, from a mathematical point of view, one has a rather rich class of K-systems. The physicist becomes expectant: is it possible to prove that N particles in a box interacting with a reasonable two-body force (say a Lennard-Jones potential) defines a K-system for suitably chosen energy surfaces?

V. HARD SPHERES IN A BOX, SINAI'S THEOREM

The only nontrivial case in which the answer to the above question is known is that of hard spheres in a box. For that case, Sinai announced in 1963 a proof that the flow defines a K-system on each energy surface.[24] The proof was to be contained in Part III of a sequence of papers of which the first appeared in 1961,[25] and the second in 1966.[26] Part III has not appeared but is expected in the winter of 1969.[27] A sketch of the proof has been published,[28] and it involves an extension to the case of discontinuous flows of the ideas used in Part II to prove that C-systems are K-systems.

The folklore says that the extension of Sinai's result to a large class of purely repulsive forces will not be difficult but that the inclusion of attraction will very likely result in a breakdown of ergodicity at least for some energy surfaces. The idea is that for sufficiently small energies one ought to be able to locate periodic orbits for approximating systems and use perturbation theory to locate periodic orbits and invariant sets of tori around them for the actual system. In preparing this review, I have not succeeded in locating any such result either in the published or preprint literature.

Thus while Sinai's result settles once and for all that statistical mechanics of an isolated mechanical system can go as the Founding Fathers suggested, the question of the ultimate status of the notions of ergodicity, mixing, and

K-system will depend on the outcome of studies of flows for more general two-body forces.

I hope that my review has made clear that in the last decade the mathematical theory of dynamical systems has absorbed some of the intuitive ideas of the founders of statistical mechanics and has transformed them into powerful general methods. In my opinion, this process has only begun and there is not only much to be done but much that can be done, both on the mathematical and physical sides. Even those of us who are forced by one circumstance or another to play a mostly pedagogical role can do something. We can let the students know that such books and reviews as Arnold-Avez,[3] Abraham,[3] Arnold,[17] Moser,[7] and Smale[3] exist and that such people as Sinai are alive and well and getting exciting results.

REFERENCES

1. For reviews of these fundamental problems see P. and T. Ehrenfest, The Conceptual Foundations of the Statistical Approach in Mechanics, (translated by M. J. Moravcsik), Cornell Univ. Press, 1959; G. E. Uhlenbeck and G. W. Ford, Lectures in Statistical Mechanics, Amer. Math. Soc., 1963.

2. L. Landau and E. Lifshitz, Statistical Physics, Vol. 5, Course of Theoretical Physics, (translated by E. Peierls and R. F. Peierls), Addison-Wesley, 1958, p. ix, say

"We do not share the view, which one encounters sometimes, that statistical physics is the least well-founded branch of theoretical physics (as regards its basic principles). We believe that the difficulties are created artificially, because the problems are often not stated sufficiently rationally. If one talks from the very beginning about the statistical distribution for small parts of a system (sub-systems) and not for a closed system as a whole, then one avoids the whole question of the ergodic or similar hypotheses, which are not really essential for physical statistics."

3. H. Poincaré, *Méthodes Nouvelles de la Méchanique Céleste.* More recent books incorporating Poincaré's point of view are not common. See, however, R. Abraham, *Foundation of Mechanics*, Benjamin, 1967; V. I. Arnold and A. Avez, *Ergodic Problems of Classical Mechanics*, Benjamin, 1968; S. Sternberg, *Celestial Mechanics* I, II, Benjamin, 1969. See also the review S. Smale, *Differentiable Dynamical Systems*, *Bull. Amer. Math. Soc.*, *73*,747 (1967).

   I have borrowed freely from these excellent accounts in preparing this talk. In particular, the striking pictures illustrating the mixing property (Figs. 3 and 4), asymptotes and horospheres (Fig. 5), and stability (Fig. 6) are adapted from Arnold-Avez.

4. See R. Abraham, Ref. 3, p. 158, for a further development.

5. The existence of the stable, unstable and center manifolds is proved in R. Abraham and J. Robbin, *Transversal Mappings and Flows*, Benjamin, 1968, Chap. VI.

6. A. Liapounov, *Ann. Math. Stud.*, No. 17.

7. J. Moser, *Amer. Math. Soc. Mem.*, No. 81, 1968.

8. See G. D. Birkhoff, *Amer. Math. Soc. Colloq. Publ.*, Vol. IX, 1927, Chap. VI.

9. See P. R. Halmos, *Lectures on Ergodic Theory*, Chelsea, 1956.

10. Ergodic flows are sometimes called metrically transitive. This is to distinguish them from topologically transitive flows; these are continuous flows such that at least one orbit is dense. They have also to be distinguished from minimal flows which are such that every orbit is dense.

11. Physicists ought to read the paper E. Hopf, J. Math. Phys., 13, 51 (1934)

12. J. Hadamard, J. Math. Pures Appl., 27 (1898).

13. E. Artin, Math. Abh., 3 170 (1924).

14. See Arnold-Avez, Appendix 27, for outline of proof and further references.

15. N. S. Krylov, Works on the Foundation of Statistical Physics (in Russian), Moscow, 1950.

16. Kolmogorov's address appears in English translation as Appendix D of Abraham's book.[3]

17. See Russ. Math. Surv., 18, 85 (1963); Russ. Math. Surv., 18, 9 (1963); and also the bibliography of Arnold and Avez.

18. J. Moser, Nach. Akad. Wiss. Gött., No. 1, 1962.

19. V. I. Arnold, Proc. Int. Cong. Math., Moscow, (1966) pp. 887-892.

20. M. Hénon and C. Heiles, Astron. J., 69, 73 (1964).

21. See, for example, A. Khinchin, Mathematical Foundations of Information Theory, Dover, 1957; P. Billingsley, Ergodic Theory and Information Theory, Wiley, 1965.

22. On the face of it, the K-S entropy of a flow has nothing to do with the entropy of an equilibrium state. For that reason I have coined the term K-S entropy. In the mathematical literature it is called entropy to the confusion of physicists. There are situations in which in fact it is equal to the ordinary entropy.

23. Ja. Sinai, Amer. Math. Soc. Trans., 68 (2), 34 (1968); see also the exposition in Chap. III of Arnold-Avez.

24. Ja. Sinai, Sov. Mat. Dokl., 4, 1818 (1963).

25. Ja. Sinai, Amer. Math. Soc. Trans., 39 (2), 83 (1964).

26. Ja. Sinai, Amer. Math. Soc. Trans., 68 (2), 34 (1968).

27. Ja. Sinai, private communication.

28. Ja. Sinai, in Statistical Mechanics: Foundations and Applications, IUPAP meeting, Copenhagen, 1966, pp. 559-573.

THE GENERALIZATION OF THE BOLTZMANN EQUATION TO HIGHER DENSITIES

E. G. D. Cohen

The Rockefeller University
New York, New York

At this moment there is no reason to doubt that the Boltzmann equation describes adequately the approach of dilute gases with short range intermolecular forces to thermal equilibrium. In particular the equation predicts correctly the density and temperature dependence of the transport coefficients: the viscosity and heat conductivity, of such a gas.[1]

For a similar description of a moderately dense gas, where in addition to binary collisions also triple and higher order collisions take place, a generalization of the Boltzmann equation to higher densities is necessary. The main content of my talk is a short review of the present situation in this respect. However, before doing so it seems useful to me to remind you very briefly of the present situation for moderately dense gases in equilibrium. Here the theory is well founded on the ensemble theory of Gibbs. One is interested in the density and temperature dependence of the thermodynamic quantities and the distribution functions of the gas.[2]

The density dependence of the thermodynamic quantities was first derived by Mayer in 1938 and is given by a power series in the density, in agreement with a much earlier empirical representation by Kamerlingh Onnes in 1911 by a power series in the reciprocal volume. Thus, for instance, for pairwise additive intermolecular forces, the pressure p is given by:[2]

$$\frac{p(n, T)}{nkT} = 1 - \sum_{\nu=1}^{\infty} \frac{\nu}{\nu+1} \beta_\nu(T) n^\nu \tag{1}$$

Here n is the density, T the absolute temperature. The β_ν are directly related to the virial coefficients and given by:

$$\beta_1(T) = \lim_{V\to\infty} \frac{1}{V} \int_V d\vec{r}_1 \int_V d\vec{r}_2 \; f(r_{12}) \tag{2}$$

$$\beta_2(T) = \lim_{V\to\infty} \frac{1}{V} \int d\vec{r}_1 \int d\vec{r}_2 \int d\vec{r}_3 \; f(r_{12}) f(r_{13}) f(r_{23}) \tag{3}$$

etc., where $f(r)$ depends on the intermolecular potential $\phi(r)$, which is assumed to be spherically symmetric:

$$f(r) = e^{-\frac{\phi(r)}{kT}} - 1$$

The νth virial coefficient $B_\nu(T) = -(\frac{\nu - 1}{\nu})\beta_{\nu-1}$.

At the present moment $B_2(T)$ can be determined experimentally to a few percent accuracy, $B_3(T)$ only to about 15% accuracy. In Figs. 1 and 2, the theoretical expressions for $B_2(T)$ and $B_3(T)$ for a 12-6 Lennard-Jones potential:

$$\phi(r) = 4\epsilon\left[\left(\frac{\sigma}{r}\right)^{12} - \left(\frac{\sigma}{r}\right)^{6}\right] \tag{5}$$

are compared with experiment.

Until now, in addition to $B_2(T)$ and $B_3(T)$, only the fourth[3,4,5] and fifth[4,6] virial coefficients, $B_4(T)$ and $B_5(T)$, have been computed for the potential (5), but no direct comparison with experiment can be made, as no reliable experimental determination of these coefficients has been reported. For hard elastic spheres seven virial coefficients have been computed.[7]

In a similar fashion as for the thermodynamic quantities, density expansions can be derived for the distribution functions of the gas.[2] For the pair distribution function $f_2(r)$ in configuration space de Boer and Mayer and Montroll derived the expansion,

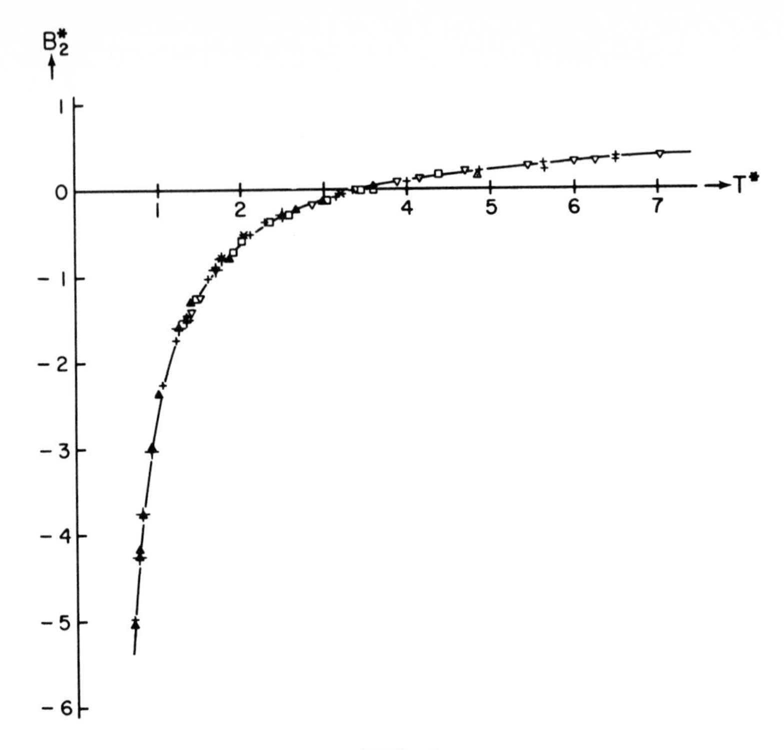

FIG. 1

Plotted is B_2^* versus T^*, where the reduced second virial coefficient $B_2^* = B_2/\frac{2}{3}\pi N\sigma^3$ and the reduced temperature $T^* = kT/\epsilon$. The reductions are carried out with the parameters ϵ and σ of the intermolecular potential field (5); N is Avogadro's number. The drawn line is the theoretical curve; the points represent experimental results for a number of substances (□ Xe, + Ar, ∇ N_2, ▲ Kr, * O_2). The values of ϵ and σ in Eq. (5) used to reduce the experimental results, are determined for each substance by choosing those values that give a best fit to the theoretical curve. According to the law of corresponding states, as formulated by de Boer[2b], B_2^* should be a universal function of T^* for all substances for which the intermolecular potential field can be represented by Eq. (5). The point where B_2 vanishes is called the Boyle point.

$$f_2(r_{12}) = n^2 g_2(r_{12}) \tag{6}$$

where the radial distribution function $g_2(r_{12})$ is given by

$$g_2(r_{12}) = \sum_{\nu=0}^{\infty} g_2^{(\nu)}(r_{12})\, n^\nu \tag{7}$$

with

$$g_2^{(o)}(r_{12}) = e^{-\frac{\phi(r_{12})}{kT}} \tag{7a}$$

$$g_2^{(1)}(r_{12}) = e^{-\frac{\phi(r_{12})}{kT}} \int d\vec{r}_3\, f(r_{13})\, f(r_{23}) =$$

$$\int d\vec{r}_3 \left[e^{-\frac{\phi(r_{12})+\phi(r_{13})+\phi(r_{23})}{kT}} - e^{-\frac{\phi(r_{12})+\phi(r_{13})}{kT}} - e^{-\frac{\phi(r_{12})+\phi(r_{23})}{kT}} + e^{-\frac{\phi(r_{12})}{kT}} \right] \tag{7b}$$

etc.

We remark that the cluster property of the $f(r)$ -- which is a consequence of the product property of the Boltzmann factor $e^{-\phi(r)/kT}$ -- ensures that the integrals in Eqs. (2), (3), and (7b) are all finite. The expression for $g_2^{(1)}(r_{12})$ given by the last integral on the right-hand side of Eq. (7b) is also valid for nonadditive intermolecular forces if one interprets $\phi(r_{12})+\phi(r_{13})+\phi(r_{23})$ as the three particle

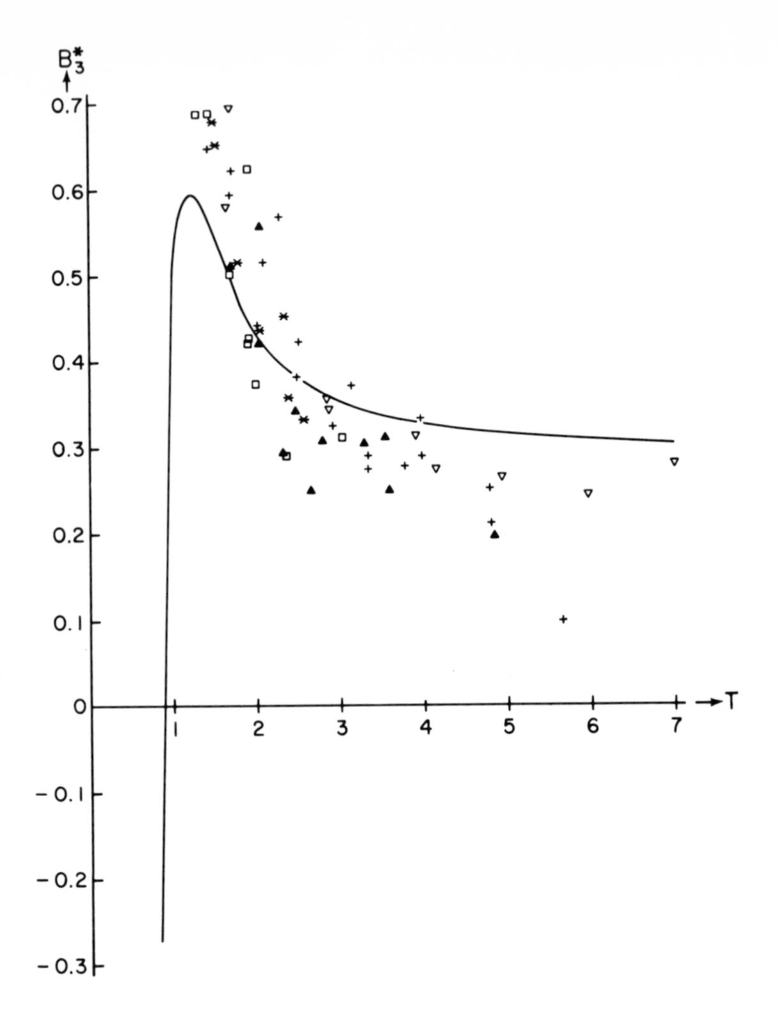

FIG. 2

Plotted is B_3^* versus T^*, where the reduced third virial coefficient $B_2^* = B_2 \neq \frac{2}{3}\pi N\sigma^3$ and the reduced temperature $T^* = kT/\epsilon$. The reductions are carried out with the parameters ϵ and σ of the intermolecular potential field, determined from $B_2^*(T^*)$; N is Avogadro's number. The drawn line is the theoretical curve; the points represent experimental results for a number of substances potential energy and is useful for a comparison with the nonequilibrium case.

Up till now, in addition to $g_2^{(o)}(r_{12})$, $g_2^{(1)}(r_{12})$, $g_2^{(2)}(r_{12})$, $g_2^{(3)}(r_{12})$ have been determined theoretically for the potential (5)[8,5,6] as well as for hard elastic spheres.[9]

In recent years several rigorous theorems have been proved for these density expansions, in particular concerning their range of convergence.[10]

In order to avoid extra complications I shall discuss the nonequilibrium theory for molecules with a pairwise additive intermolecular potential, where the potential is monotonically repulsive and of finite range r_o.

The gas can then be adequately described by the single particle and pair distribution functions $f_1(\vec{r}_1\, \vec{p}_1\, ; t)$ and $f_2(\vec{r}_1\, \vec{p}_1\, \vec{r}_2\, \vec{p}_2\, ; t)$ respectively.

(□Xe, + Ar, ∇N_2, ▲Kr, * O_2). According to the law of corresponding states, as formulated by de Boer[2b], B_3^* should be a universal function of T^* for all substances for which the intermolecular potential field can be represented by Eq. (5). The scatter of the experimental points is probably mainly due to experimental inaccuracies, but may also be caused by nonvalidity of the law of corresponding states.

For a moderately dense gas two equations can be derived for f_1 and f_2, which together describe the time evolution of f_1 and f_2, and therefore of the gas, for a given initial state.

One equation connecting f_1 and f_2 can be obtained directly from the Liouville equation by integration and requiring the vanishing of the distribution functions for $\vec{r} \to \infty$ as well as $\vec{p} \to \infty$.

The second equation expresses f_2 in a formal density expansion and can be derived in a way analogous to Eq. (7), by using appropriate nonequilibrium generalizations of the cluster functions introduced by Uhlenbeck, Kahn, and de Boer in the equilibrium theory of moderately dense gases. Due to the finite range r_o of the intermolecular potential, f_2 is only needed and considered here for interparticle distances $r_{12} = |\vec{r}_1 - \vec{r}_2| \leq r_o$.

Introducing $F_1(1;t) = f_1(\vec{r}_1\vec{p}_1;t)/n$ and $F_2(1,2;t) = f_2(\vec{r}_1\vec{p}_1, \vec{r}_2\vec{p}_2;t)/n^2$ one has in the thermodynamic limit:[11]

$$\frac{\partial F_1(1;t)}{\partial t} = -\frac{\vec{p}_1}{m} \cdot \frac{\partial F_1}{\partial \vec{r}_1} + n \int d\vec{r}_2 \int d\vec{p}_2 \, \Theta_{12} F_2(1,2;t) \tag{8}$$

where

$$\Theta_{12} = \frac{\partial \phi(r_{12})}{\partial \vec{r}_{12}} \cdot \left(\frac{\partial}{\partial \vec{p}} - \frac{\partial}{\partial \vec{p}_2} \right) \tag{8a}$$

and

$$F_2(1, 2; t) = \sum_{\nu=1}^{\infty} \tilde{F}_2^{(\nu)}(1, 2; t) n^{\nu} \tag{9}$$

where

$$\tilde{F}_2^{(o)}(1, 2; t) = \tilde{\mathscr{S}}_t(12) F_1(1; t) F_1(2; t) \tag{9a}$$

$$\tilde{F}_2^{(1)}(1, 2; t) = \int d3 [\tilde{\mathscr{S}}_t(123) - \tilde{\mathscr{S}}_t(12)\tilde{\mathscr{S}}_t(13) - \tilde{\mathscr{S}}_t(12)\tilde{\mathscr{S}}_t(23) + + \tilde{\mathscr{S}}_t(12)] \prod_{i=1}^{3} F_1(i; t) \tag{9b}$$

etc., with

$$\tilde{\mathscr{S}}_t(12\ldots s) = S_{-t}(12\ldots s) a_s(12\ldots s; o) \prod_{i=1}^{s} S_t(i) \tag{10a}$$

Here the operator $S_t(12\ldots s) = \exp it L_s(12\ldots s)$ transforms the phases of the particles $1, 2, \ldots, s$ into those at a time t later if these particles move under their mutual interaction only. $L_s(12\ldots s) = -i\{\ , H_s\}$, where $\{\ \}$ are Poisson brackets and H_s is the Hamilton function of the particles $12\ldots s$, so that $L_s(1\ldots s)$ is the Liouville operator for the particles $12\ldots s$. The $a_s(12\ldots s; o)$, characterize the correlations between the s particles $1, 2, \ldots s$ at $t = 0$ and can be expressed in terms of $b_s(12\ldots s; o)$ defined by:

$$F_s(12\ldots s; o) = b_s(12\ldots s; o) \prod_{i=1}^{s} F_1(i; o). \tag{10b}$$

Thus f. i. the relation between b_2 and a_s is given by Eq. (7), if one replaces on the left hand side g_2 by b_2 and on the right hand side in $g_2^{(\nu)}$ the Boltzmann factor for s particles by a_s.

We shall assume that the b_s have a product property with respect to r_o and are equal to one if all s particles are outside each other's range of interaction at $t = 0$. The same properties follow then for the a_s.

The structure of the F_2-expansion Eq. (9) is similar to that of the g_2-expansion Eq. (7): the $\tilde{F}_\nu^{(2)}$ be considered as a "dynamical" generalization of the $g_2^{(\nu)}$, in that the Boltzmann factor relating to s particles is replaced by the dynamical s-particle operator $\tilde{\mathscr{S}}_t(12\ldots s)$, while $F_1(t)$ replaces n.

In the Eqs. (8) and (9) only those contributions to $\partial F_1/\partial t$ and to F_2 have been neglected that vanish in the thermodynamic limit. Under this restriction they can be considered exact consequences of the Liouville equation. The Eqs. (8) and (9) form a convenient starting point for a discussion of the approach of a moderately dense gas to thermal equilibrium.

In order to describe the evolution in time of the gas to thermal equilibrium, these equations have to be taken in a coarse-grained sense. Thus Bogolubov postulated in 1946 on physical grounds that after a time $t \gg t_{coll}$, the average duration of a binary collision, the initial state of the gas could be forgotten and f_2 could be assumed to be a time independent functional of f_1[12]. In terms of Eqs. (8) and (9) this would imply that:

a) for $t \gg t_{coll}$, the $a_s(12 \ldots s; 0)$ can be set equal to one;

b) for $t \gg t_{coll}$, the t in $\tilde{\mathscr{S}}_t$ can be replaced by ∞;

c) for $t \to \infty$, $F_1(t)$ approaches the Maxwell distribution function, and the $F_2^{(\nu)}$ approach the $g_2^{(\nu)}$ of Eq. (7).

A study of the first three terms on the right-hand side of the Eq. (9) has indeed shown that:[11]

a) for $t \gg t_{coll}$, a_2, a_3, and a_4 can be set equal to one in $\tilde{F}_2^{(o)}$, $\tilde{F}_2^{(1)}$, and $\tilde{F}_2^{(2)}$ respectively and

b) for $t \gg t_{coll}$, $\tilde{\mathscr{S}}_t(1 \ldots s)$ can be replaced by

$$\mathscr{S}_\infty(1 \ldots s) = \lim_{t \to \infty} \mathscr{S}_t(1 \ldots s) = \lim_{t \to \infty} S_{-t}(1 \ldots s) \prod_{i=1}^{\infty} S_t(i)$$

in $\tilde{F}_2^{(o)}$ and $\tilde{F}_1^{(1)}$, without causing any difficulty. These results have been obtained using product properties of the a_s as well as of the $\mathscr{S}_t(1 \ldots s)$. Thus, for $t \gg t_{coll}$, one can transform $\tilde{F}_2^{(o)}$ and $\tilde{F}_2^{(1)}$ into $F_2^{(o)}$ and $F_2^{(1)}$ given by:

$$F_2^{(o)}(12\,;t) = \mathscr{S}_\infty(12)\, F_1(1\,;t)\, F_1(2\,;t) \tag{11a}$$

$$F_2^{(1)}(12\,;t) = \tag{11b}$$

$$= \int d3\, [\mathscr{S}_\infty(123) - \mathscr{S}_\infty(12)\,\mathscr{S}_\infty(13) - \mathscr{S}_\infty(12)\,\mathscr{S}_\infty(23) + \mathscr{S}_\infty(12)] \prod_{i=1}^{3} F_1(i;t)$$

respectively.

Substituting $F_2^{(o)}$ for F_2 into Eq. (8) yields for F_1 the Boltzmann equation plus a correction which takes into account the difference in position of the two molecules engaged in a binary collision, which is neglected in the Boltzmann equation.[12] Using in addition $F_2^{(1)}$ in Eq. (8) gives a correction to the Boltzmann equation due to triple collisions.

c) $\tilde{F}_2^{(\nu)}$ reduces to $g_2^{(\nu)}$ if $F_1(t)$ is replaced by the Maxwell distribution function.[12a, c, 13]

I remark that one has <u>not</u> been able to prove the approach of $F_1(t)$ for $t \to \infty$ to the Maxwell distribution function other than through binary collisions. Not even the approach of a gas to thermal equilibrium through binary <u>and</u> triple collisions, i.e., using $F_2 = F_2^{(o)} + F_2^{(1)}$ in Eq. (8) has been proven at this moment.

However, a much more serious difficulty is that $\tilde{F}_2^{(2)}$, $\tilde{F}_2^{(3)}$, etc., have been shown to contain contributions that grow with time, so that these terms diverge in the limit $t \to \infty$. This has been found independently by Weinstock, Frieman, and Goldman, and Dorfman and Cohen.[14] In fact, for $t >> t_{coll}$, $\tilde{F}_2^{(2)} \sim \ln(t/t_{coll})$ and $\tilde{F}_2^{(\nu)} \sim (t/t_{coll})^{\nu-2}$, so that the expansion of Eq. (9) for F_2 cannot be used to obtain the behavior of F_2 for $t >> t_{coll}$ for a moderately dense gas and Bogolubov's postulates cannot be correct.

In order to obtain an expansion of F_2 usable for $t >> t_{coll}$, one has to rearrange Eq. (9). Before I do this, let me first analyze the origin of the divergent contributions to $F_2^{(\nu)}$ $(\nu \geq 2)$. The contributions to $\tilde{F}_2^{(1)}$ (and to $\tilde{F}_2^{(\nu)}$ in general) can be classified according to the dynamical events (collisions) that occur between the particles 123 (or $123 \ldots \nu+2$ in general) between $t = 0$ and t. In addition to a genuine triple collision (or a genuine ν-particle collision), extended collision sequences consisting of three (in general $(\nu+2)$) or more successive binary collisions contribute to $\tilde{F}_2^{(1)}$ (or $\tilde{F}_2^{(\nu)}$) (cf. Fig. 3). It is these extended binary collision sequences that cause the divergences mentioned above, since a) there is no restriction on the time available for the sequence of binary collisions to take place if $t \to \infty$ and b) the phase space of particle 3 (or $3, 4 \ldots \nu+2$) for fixed phases of the particles 1 and 2 (with $r_{12} \leq r_o$) for 3 (or $(\nu+2)$) successive binary collisions to occur grows $\sim \ln(t/t_{coll})$ or $(\sim (t/t_{coll})^{\nu-4})$ for $t >> t_{coll}$. Fig. 4 shows a few collision sequences that lead to divergent contributions in $\tilde{F}_2^{(2)}$.

We remark that in addition to real binary collisions, also imaginary binary collisions have to be considered, where the

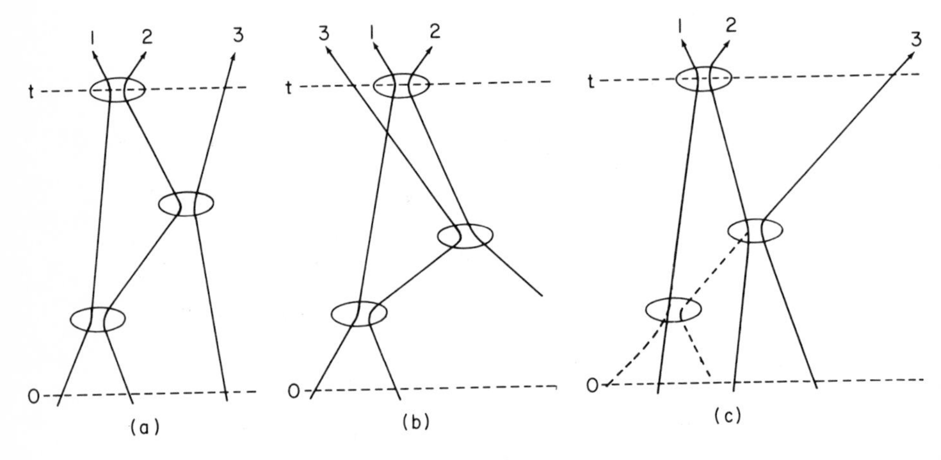

FIG. 3

Binary collision sequences contributing to $\tilde{F}_2^{(1)}$. Because of the S_{-t} operators occurring in Eq. (9), one is interested in the dynamical events as they occur moving <u>backwards</u> in time from t to 0. (ij) denotes a real-, $(\overline{ij})$ an imaginary collision between the particles i and j. (a) three real binary collisions: (12) (23) (12) (a recollision between the particles 1 and 2). (b) three real binary collisions: (12) (23) (13). (c) two real and one imaginary binary collision: on moving backwards in time particle 3 would have collided with particle 1, but for the intervention of particle 2: (12) (23) $(\overline{13})$.

particles would have collided, but for the interference of a third (cf. Fig. 3c). It is generally assumed that the collision sequences discussed here constitute, for the potential considered here, the class of most divergent contributions in each order of n. One would hope that by resumming in the expansion Eq. (9) all these most divergent contributions, an expansion with only finite or at least less divergent terms can be obtained.

In order to carry out such a resummation of Eq. (9), one makes a binary collision expansion of the $\mathscr{S}_t(1 \ldots s)$-operator operators.

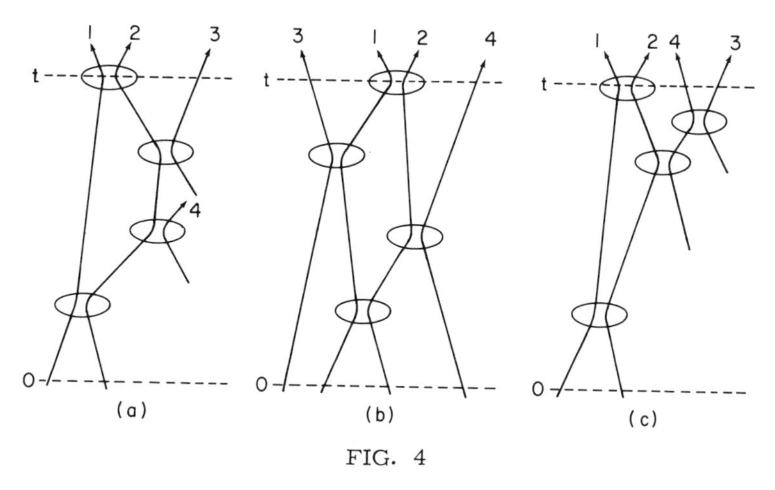

FIG. 4

Binary collision sequences contributing to $\tilde{F}_2^{(2)}$; four real binary collisions (all recollisions between the particles 1 and 2): (a): (12) (23) (24) (12). (b): (12) (13) (24) (12). (c): (12) (34) (24) (12).

The binary collision expansion enables one in a systematic way to isolate the most divergent contributions in the limit of $t \to \infty$ in each order of the density and effect a partial resummation of these contributions. A disadvantage is that also finite contributions, for instance the genuine n-body collisions, are expanded in a series of binary collisions. The binary collision expansion was introduced by Yang and Lee[15] for the treatment of the hard sphere Bose gas in equilibrium, and adapted to nonequilibrium statistical mechanics by Weinstock and Zwanzig.[16] For the purpose of resummation, it was first used by Weinstock and Kawasaki and Oppenheim.[17]

Using the binary collision expansion, all most divergent contributions in each order of n can be combined in the triple collision term, effecting two important changes: a) introducing a "damping" of the paths of the colliding particles 1 and 2, restricting each path essentially to the mean free path; b) changing the time dependence of the F_1, which were all at the same time t, to times t_i before t. Thus, after rearrangement, one has $(r_{12} \leq r_o)$:[18]

$$F_2(12\,;t) = F_2^{(o)}(12\,;t) + n\overline{F}_2^{(1)}(12\,;t) \tag{12}$$

where $F_2^{(o)}$ is given by Eq. (11a) and $\overline{F}_2^{(1)}$ by (cf. Fig. 5):

$$\overline{F}_2^{(1)}(12\,;t) = n \int_0^t dt_1 \int_0^t dt_2\, \mathscr{S}_\infty(12) \prod_{i=1}^{2} S_{-(t-t_1)}(i)$$

$$\int d3\, \lambda_3(12)\, F_1(3\,;t_1) \prod_{i=1}^{2} S_{-t_1}(i)$$

$$T \exp \left[n \int_{t_2}^{t_1} d\tau \prod_{i=1}^{2} S_\tau(i) \int d4\, \lambda_4(12)\, F_1(4\,;\tau) \prod_{i=1}^{2} S_{-\tau}(i)\right] \tag{12a}$$

$$\prod_{i=1}^{2} S_{t_2}(i)\, \Theta_{12}\, \mathscr{S}_\infty(12)\, F_1(1\,;t_2)\, F_1(2\,;t_2)$$

+ etc.

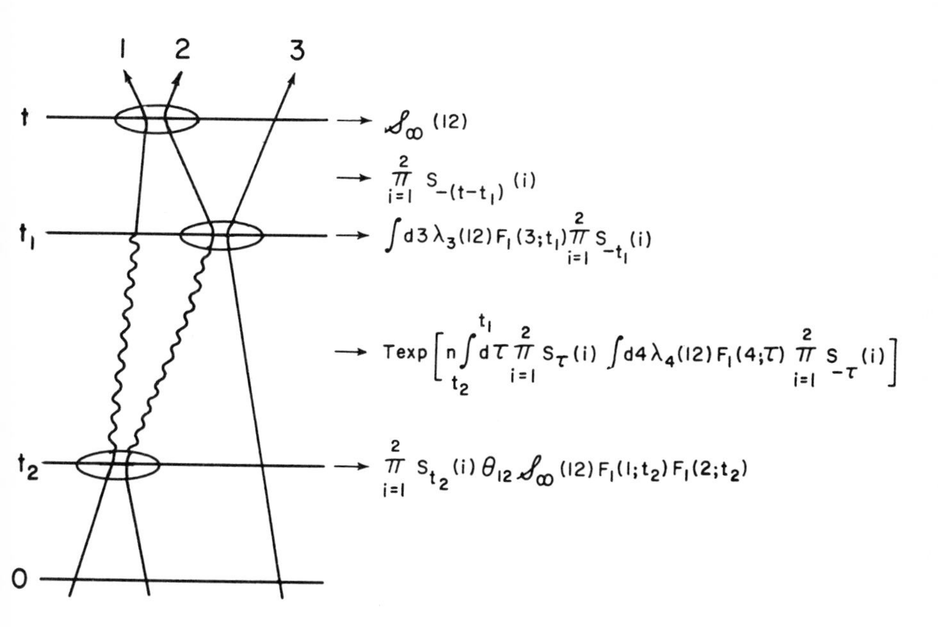

FIG. 5

Recollision event (12) (23) (12) after resummation (cf. Fig. 3a). The presence of the damping operator between t_1 and t_2 is schematically indicated by representing the paths of the particles 1 and 2 by wavy lines during that time interval.

Here, etc. includes the contributions from all those configurations of the particles 1, 2, and 3 where in addition to $r_{12} \leq r_o$ either $r_{13} \leq r_o$ or $r_{23} \leq r_o$ or both, i.e., from configurations where either two pairs of particles or all three particles overlap each other. These configurations did not contribute to the divergences and are of no special interest here.

$$\lambda_i(12) = \Theta_{1i} \mathscr{S}_\infty(1i)(1 + P_{1i}) + \Theta_{2i} \mathscr{S}_\infty(2i)(1+P_{2i})$$

where P_{ij} is the permutation operator on i and j ; T indicates that the exponential that follows it has to be interpreted as a time-ordered product.

The exponential, whose exponent is related to the Boltzmann collision operator, acts as a damping and incorporates the effect of successive binary collisions of the particles 1 and 2 with other particles in the gas. It introduces a new time, the average time to transverse a mean free path, t_{mfp}. A different derivation of this damping exponential has been given by Frieman and Goldman.[19]

One can show that after this resummation, the new four body term is convergent in the limit $t \to \infty$, the events that lead to the $\ln(t/t_{coll})$ behavior having been absorbed in the damping and the time changes in $\overline{F}_2^{(1)}$. However, the higher order terms still contain contributions that grow with time, though these contributions grow, in each order of n, less rapidly with time than those that were eliminated by the resummation. One could hope that these contributions can be incorporated into a damping and appropriate time changes in the F_1 in the four body term and so on, ultimately leading to an expansion without divergences altogether. This program has not been carried out, mainly because a complete classification of sequences of collisions according to their behavior for $t >> t_{coll}$ has not been carried out for the ν-body term with $\nu \geq 4$, in Eq. (9).

The Eq. (12) when introduced for f_2 into Eq. (8), leads to a nonlinear evolution equation for $F_1(t)$. This equation has a very different structure than the Boltzmann equation, in that $\partial F_1(t)/\partial t$ depends not only on $F_1(t)$ at the same time t but on $F_1(t)$ at all t_i before t as well. In view of the damping operator, one would expect that this dependence would actually not go back more than a few mean free times t_{mfp}.

In spite of the peculiar structure of $\overline{F}_2^{(1)}$, the behavior of the gas near thermal equilibrium or the approach of the gas to equilibrium for $t >> t_{mfp}$, can be obtained from the

Eqs. (8) and (12) for F_1 by linearizing around local equilibrium. In doing so, one obtains a linearized equation for F_1 from which the linearized Euler- and Navier-Stokes equations of hydrodynamics can be derived with explicit expressions for the transport coefficients, the viscosity and the heat condictivity of the gas, in terms of the intermolecular forces.[18]

These expressions can be shown to be identical to those that can be obtained from a density expansion of the Green-Kubo time correlation function expressions for the transport coefficients.[20]

Including the effects of quadruple collisions, one obtains

$$\eta(\rho,T) = \eta_o(T) + \rho\eta_1(T) + \rho^2 \ln\rho\,\eta_2'(T) + \rho^2\eta_2(T) \qquad (13a)$$

$$\lambda(\rho,T) = \lambda_o(T) + \rho\lambda_1(T) + \rho^2 \ln\rho\,\lambda_2'(T) + \rho^2\lambda_2(T) \qquad (13b)$$

where $\rho = n\,r_o^3$.

Here $\eta_o(T)$ and $\lambda_o(T)$ are contributions to $\eta(T)$ and $\lambda(T)$ respectively due to binary collisions and have been obtained by Chapman and Enskog from the Boltzmann equation.[1] $\eta_o(T)$ and $\lambda_o(T)$ are plotted in Fig. 6. They can be determined from experiment to a few percent accuracy.

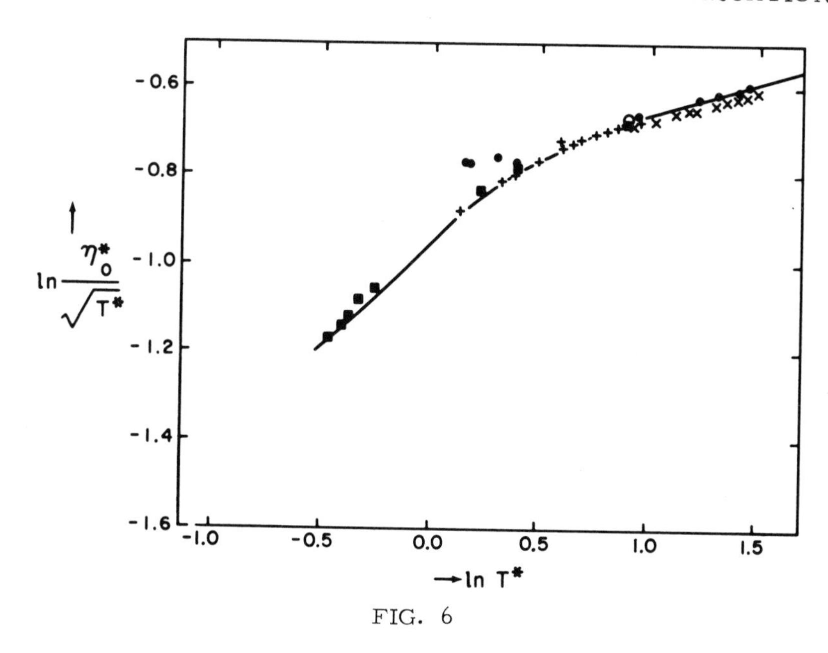

FIG. 6

Plotted is $\ln \eta_0^*/\sqrt{T^*}$ versus $\ln T^*$, where the reduced viscosity $\eta_0^* = \eta_0 \sigma^2/(m\epsilon)^{1/2}$ and the reduced temperature $T^* = kT/\epsilon$. The reductions are carried out with the parameters ϵ and σ of the intermolecular potential field (5) determined from $B_2^*(T^*)$, as well as with the molecular mass m. The drawn line is the theoretical curve; the points represent the experimental results for a number of substances (O H_2, ■ D_2, x Ne, + Ar, · He). According to the law of corresponding states, as formulated by de Boer[2b,1a,1d] $\ln \eta_0^*/\sqrt{T^*}$ should be a universal function of $\ln T^*$ for all substances for which the intermolecular potential field can be represented by Eq. (5). As $\ln \lambda_0^*/\sqrt{T^*} \cong \ln \lambda_0^*/\sqrt{T^*}$ + ln 15/4, where $\lambda_0^* = \lambda_0 \sigma^2 (m/\epsilon)^{1/2}/k$, the plot of $\ln \lambda_0^*/\sqrt{T^*}$ versus $\ln T^*$ is essentially the same as that given in Fig. 6.

$\eta_1(T)$ and $\lambda_1(T)$ contain in addition to negative contributions η_1^κ, λ_1^κ from triple collisions due to $\overline{F}_2^{(1)}$, positive contributions η_1^ϕ, λ_1^ϕ, arising from the difference in position of two colliding molecules in a binary collision, due to $F_2^{(o)}$. These contributions represent the collisional transfer of momentum and energy through the intermolecular potential field, as opposed to the kinetic transfer of momentum and energy through free flight, incorporated in η_1^κ and λ_1^κ. We remark that the density dependence of the transport coefficients is <u>not</u> given by a power series in the density but contains a term that depends logarithmically on the density. This term is directly related to the logarithmic behavior $\sim \ln(t/t_{coll})$ for $t >> t_{coll}$ of the four particle term $\tilde{F}_2^{(2)}$ in Eq. (9). In fact, the occurrence of a $\rho^2 \ln \rho$-term can be intuitively understood if one assumes that the introduction of the damping in Eq. (12) effectively cuts off all t at t_{mfp} so that $\rho^2 \ln(t/t_{coll})$ becomes $\rho^2 \ln(t_{mfp}/t_{coll}) \sim \rho^2 \ln \rho$ as $t_{mfp} \sim 1/nr_o^2$. Dorfman has been able to show that:

a) $\eta_1(T)$ is the same whether obtained from $F_2^{(o)} + F_2^{(1)}$ or from $F_2^{(o)} + \overline{F}_2^{(1)}$, i. e., $\eta_1(T)$ is unaffected by the resummation. Expressions for $\eta_1(T)$ and $\lambda_1(T)$ in terms of the intermolecular potential were first obtained by Choh and Uhlenbeck, using $F_2^{(o)} + F_2^{(1)}$ for F_2 in Eq. (8).[12c,d]

b) The coefficient of the logarithmically divergent contribution of $0(\rho^2)$ to $\eta(T)$ and $\lambda(T)$ that would be obtained if $F_2^{(2)}$ instead of the resummed expression $\overline{F}_2^{(2)}$ were used in its computation, is just η_2' or λ_2' respectively, i. e., for the computation of $\eta_2'(T)$ and $\lambda_2'(T)$ the resummation does not have to be carried out. This is not true of $\eta_2(T)$ or $\lambda_2(T)$. $F_2^{(2)}$ is obtained from $\tilde{F}_2^{(2)}$ by setting all $a_s = 1$ and replacing $\mathscr{S}_t$ by $\mathscr{S}_\infty$ (cf. Eqs. 11a, b).

Up till now only $\eta_1(T)$ and $\lambda_1(T)$ have been computed and only for hard elastic spheres. This formidable calculation has been carried out by Sengers.[21] The main result of interest here is that 95% of the contributions to η_1^κ and λ_1^κ come from configurations of the particles 1, 2, and 3 where all three particles are within each other's range of interaction i. e., from the same configurations that contribute to the coefficient $g_2^{(1)}(r_{12})$ in thermal equilibrium. This implies that the contributions to η_1^κ and λ_1^κ overwhelmingly come from excluded volume contributions and not from momentum correlations as f. i. due to recollisions, i. e., repeated collisions of the same particles (cf. Fig. 3a). One would be inclined to believe that also for more general intermolecular potentials, spatial correlations are much more important than momentum correlations in determining $\eta_1^{\kappa}(T)$ and $\lambda_1^{\kappa}(T)$, which would greatly facilitate their computation. An approximate calculation of $\eta_1(T)$ and $\lambda_1(T)$ for a 12-6 Lennard-Jones potential has been made by Curtiss c. s.: Snider and Curtiss[22] computed η_1^{ϕ} and λ_1^{ϕ} from the exact expressions based on $F_2^{(o)}$; Hoffman and Curtiss[23] used very simplified expressions for the computation of η_1^{κ} and λ_1^{κ}, not based on $F_2^{(1)}$.

Their results are plotted in Figs. 7 and 8 together with experimental values for $\eta_1(T)$ and $\lambda_1(T)$ derived from experiment by Hanley, McCarty, and Sengers on the basis of (13 a, b) with an accuracy of about 15%.[24] I must say, however, that as of this moment no conclusive experimental evidence for the existence of a logarithmic term in the density dependence of η and λ has been obtained.[25] I also remark that the absolute values for $\eta_1(T)$ and $\lambda_1(T)$ are small, due to a partial cancellation of η_1^{κ}, λ_1^{κ} and η_1^{ϕ}, λ_1^{ϕ} respectively. There are two zero points for $\eta_1(T)$ which could be called Boltzmann points in analogy with the Boyle point of $B_2(T)$ (cf. Fig. 1). In view of the restrictions on the potential, the low temperature behavior cannot be expected to be well predicted by the theory. Still, it seems to

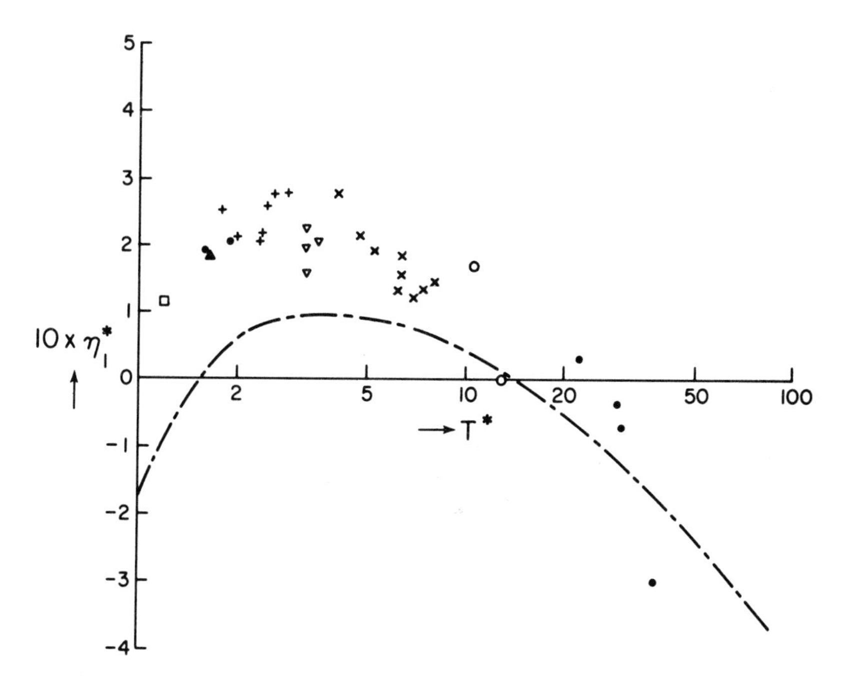

FIG. 7

Plotted is $10\,\eta_1^*$ versus T^*, where the reduced coefficient $\eta_1^* = \eta_1\sigma^2/(m\epsilon)^{1/2}$ and the reduced temperature $T^* = kT/\epsilon$. The broken line is the theoretical curve computed by Curtiss c. s.; the experimental results are represented by points (• He, x Ne, + Ar, ∇ N_2, ▲Kr, OH_2). According to the law of corresponding states, as formulated by de Boer,[2b, 1a, 1d] η_1^* should be a universal function of T^* for all substances for which the intermolecular potential field can be represented by Eq. (5). The points where η_1^* vanishes could be called Boltzmann points (cf. legend for Fig. 1). The scatter of the experimental points is probably mainly due to experimental inaccuracies but may also be caused by nonvalidity of the law of corresponding states.

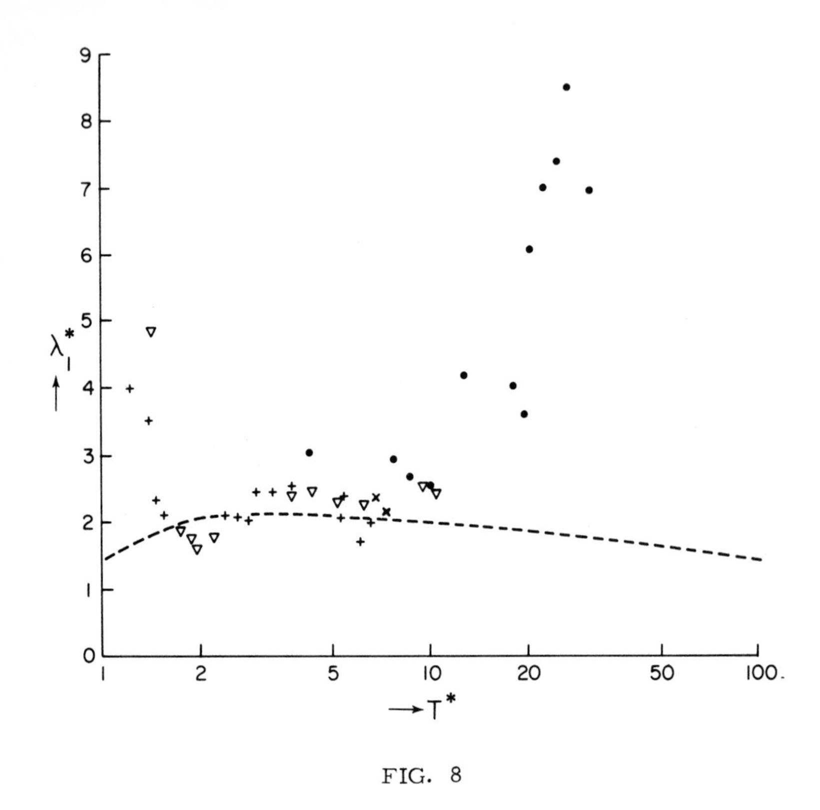

FIG. 8

Plotted is λ_1^* versus T^*, where the reduced coefficient $\lambda_1^* = \lambda_1 \sigma^2 (m/\epsilon)^{1/2}/k$ and the reduced temperature $T^* = kT/\epsilon$. The broken line is the theoretical curve computed by Curtiss c. s.; the experimental results are represented by points (• He, x Ne, + Ar, ∇N_2). According to the law of corresponding states as formulated by de Boer,[2b, 1a, 1d] λ_1^* should be a universal function of T^* for all substances for which the intermolecular potential field can be represented by Eq. (5). The scatter of the experimental points is probably mainly due to experimental inaccuracies but may also be caused by nonvalidity of the law of corresponding states.

me that $\eta_1(T)$ provides a very sensitive test of the intermolecular potential field and could even be used perhaps to obtain information of the nonadditivity of the intermolecular forces. However, as long as no calculations have been carried out for η_1^κ and λ_1^κ for a realistic intermolecular potential like Eq. (5) using the exact expressions for η_1^κ and λ_1^κ based on $F_2^{(1)}$, a real confrontation of theory and experiment is not possible.

I would like to make a few concluding remarks.

a) A major flaw of the kinetic theory of moderately dense gases is that no proof of the approach to thermal equilibrium exists if one considers triple collisions in addition to binary collisions. Not even for a spatially homogeneous gas, where F_1 depends only on $\vec{p}$ and not on $\vec{r}$, does a proof exist; in particular no H-theorem has been proven for such a gas.

b) A number of simplified models have been studied which, apart from some model dependent features, illustrate and confirm the above mentioned results.[26] In particular, the elimination of divergences by the introduction of the damping in Eq. (12a) has been verified. Only in very few cases has a rigorous result been obtained.[27,26a] Nontrivial rigorous results are hard to get, partly because one does not know the theorems to be proved. A great obstacle also seems to be the absence of a classification of all real and imaginary collisions between a given number of particles. Only for three identical hard spheres has a beginning been made.[28]

c) The same divergence difficulty as discussed above for the classical gas also seems to occur in the case of a quantum mechanical gas.[29]

d) The divergences discussed above show that a direct generalization to nonequilibrium phenomena of the virial expansion, i. e. , an expansion of the macroscopic properties of the gas in terms of properties of small isolated groups of particles, is not possible. Yet the rearrangement of the expansion (9) into (12) still leaves the idea of an expansion in terms of small groups of particles intact and only changes the expansion, in that it limits the free path of a particle to the mean free path. One could call this a modified individual particle theory, as the collective effects due to the other particles come in, in a rather trivial manner. On the other hand these effects are responsible for the occurrence of the logarithmic density dependence of η and λ .

If I may end by speculating on a possible development in the next decade, I believe that attention will be paid to the abnormal behavior in the approach to equilibrium in moderately dense systems due to more fundamental collective effects, than considered here and which have already been noticed in a number of cases. In my opinion their study may well lead to a better understanding from a microscopic point of view of the "critical slowing down" that is known to occur near a critical point for a phase transition.

The cases I have in mind are:

a) The abnormally slow diffusion found for the Ehrenfest wind-tree model - where independent point particles move through an array of squares that are randomly arranged in the plane as far as their center is concerned, but have corresponding diagonals parallel to each other - in case the squares are allowed to overlap each other. For this case the possibility of an abnormal diffusion process was predicted theoretically by Hauge and Cohen[26d] and found "experimentally" by computer experiments by Wood and Lado.[30]

b) An abnormal drop in the electron mobility in helium gas, discovered experimentally by Levine and Sanders[31] and discussed theoretically by Neustadter and Coopersmith.[32]

c) An abnormal behavior of two-dimensional systems in general, noted theoretically by Pomeau[33] and perhaps "experimentally" by computer experiments by Alder and Wainwright[34] in a system of hard discs.

In all these cases correlations between particles seem to modify in an essential way the macroscopic behavior of the system as far as the transfer of mass, momentum, and energy is concerned.

ACKNOWLEDGMENTS

I am indebted to Drs. J. M. H. Sengers-Levelt, D. Henderson, and H. J. Hanley for providing me with much of the experimental data used in this lecture.

REFERENCES

1. For the kinetic theory of dilute gases based on the Boltzmann equation see:
   a) S. Chapman and T. G. Cowling, The Mathematical Theory of Non-Uniform Gases, Cambridge Univ. Press, England, 1953.
   b) L. Waldmann, Encyclopedia of Physics (S. Flügge, ed.), Vol. XII, Springer Verlag, Berlin, 1958.
   For an introduction see:
   c) G. E. Uhlenbeck and G. W. Ford, in Lectures in Applied Mathematics, Vol. I, Amer. Math. Soc., Providence, 1963, Chaps. 4 and 6.
   d) E. G. D. Cohen, in Transport Phenomena in Fluids (H. J. M. Hanley, ed.), Marcel Dekker, New York 1969, Chap. 6.

2. For a survey, see:
   a) G. E. Uhlenbeck and G. W. Ford, in Studies in Statistical Mechanics, Vol. I, (J. de Boer and G. E. Uhlenbeck, eds.), North-Holland Publ. Co., Amsterdam, 1962, Part B, p. 123.
   b) J. de Boer, Rept. on Prog. in Phys., XII, 305 (1949).

3. J. A. Barker and J. J. Monaghan, J. Chem. Phys., 36, 2564 (1962).

4. J. A. Barker, D. J. Leonard, and A. Pompe, J. Chem. Phys., 44, 4206 (1966).

5. D. Henderson and L. Oden, Mol. Phys., 10, 405 (1966).

6. S. Kim, D. Henderson, and L. Oden, J. Chem. Phys., 45, 4030 (1966); Disc. Faraday Soc., 43, 26 (1967); Trans. Faraday Soc., 65, 2308 (1969).

7. See, F. H. Ree and W. G. Hoover, J. Chem. Phys., 46, 4181 (1967), for the seventh virial coefficient and for further references.

8. D. Henderson, Mol. Phys., 10, 73 (1965).

9. F. H. Ree, R. N. Keeler, and S. L. MacCarthy, J. Chem. Phys., 44, 3407 (1966), for $g_2^{(3)}$ and further references.

10. See D. Ruelle, Statistical Mechanics, W. A. Benjamin, New York, 1969.

11. J. R. Dorfman and E. G. D. Cohen, J. Math. Phys., 8, 282 (1967).

12. a) N. N. Bogolubov, in Studies in Statistical Mechanics, Vol. I, (J. de Boer and G. E. Uhlenbeck, eds.), North-Holland Publ. Co., Amsterdam, 1962, Part A, p. 5.
    b) G. E. Uhlenbeck and G. W. Ford, in Lectures in Applied Mathematics, Vol. I, Amer. Math. Soc., Providence, 1963, Chap. 7.
    c) S. T. Choh and G. E. Uhlenbeck, The Kinetic Theory of Dense Gases, Univ. of Michigan Rep., 1958.
    d) E. G. D. Cohen, in Fundamental Problems in Statistical Mechanics, Vol. I, North-Holland Publ. Co., Amsterdam, 1962, p. 110.

13. R. Piccirelli, J. Math. Phys., 7, 922 (1966).

14. a) J. Weinstock, Phys. Rev., 132, 454 (1963); 140 A, 460 (1965).

b) R. Goldman and E. A. Frieman, Bull. Amer. Phys. Soc., 10, 531 (1965); J. Math. Phys., 8, 1410 (1967).
c) J. R. Dorfman and E. G. D. Cohen, Phys. Letters, 16, 124 (1965); J. Math. Phys., 8, 282 (1967).

15. C. N. Yang and T. D. Lee, Phys. Rev., 113, 1165 (1969); see also A. J. F. Siegert and E. Teramoto, Phys. Rev., 110, 1212 (1958).

16. J. Weinstock, Phys. Rev., 126, 341 (1962); 132, 454, 470 (1963); R. W. Zwanzig, Phys. Rev., 129, 486 (1963); see also, J. R. Dorfman, in Lectures in Theoretical Physics, Vol. 9C, (W. E. Brittin and A. O. Barut, eds.), Gordon and Breach, New York, 1967, Vol. 9C, p. 443.

17. J. Weinstock, Phys. Rev., 140A, 460 (1965); Phys. Rev. Letters, 17, 130 (1966); K. Kawasaki and I. Oppenheim, Phys. Rev., 136A, 1519 (1964); see also L. K. Haines, J. R. Dorfman, and M. H. Ernst, Phys. Rev., 144, 207 (1966).

18. J. R. Dorfman and E. G. D. Cohen, to be published.

19. E. A. Frieman and R. Goldman, J. Math. Phys., 7, 2153 (1966); 8, 1410 (1967); see also J. R. Dorfman, Renormalised Kinetic Equations, Univ. of Maryland Tech. Note BN618, Univ. of Maryland, 1969.

20. M. H. Ernst, J. R. Dorfman, and E. G. D. Cohen, Phys. Letters, 12, 319 (1964); Physica, 31, 493 (1965); M. H. Ernst, Thesis, Univ. of Amsterdam, 1965.

21. J. V. Sengers, Phys. Fluids, 9, 1685 (1966); in Lectures in Theoretical Physics, Vol. 9C (W. E. Brittin and A. O. Barut, eds.), Gordon and Breach, New York, 1967, p. 335.

22. R. F. Snider and C. F. Curtiss, Phys. Fluids, 1, 122 (1958).

23. D. K. Hoffman and C. F. Curtiss, Phys. Fluids, 8, 890 (1965).

24. H. M. Hanley, R. D. McCarty, and J. V. Sengers, J. Chem. Phys., 50, 857 (1969).

25. J. R. Dorfman and J. V. Sengers have pointed out that a similar logarithmic term, due to similar dynamical events, is present in the expansion of the drag coefficient of a sphere in a gas stream in terms of the inverse Knudsen number.

26. a) J. L. Lebowitz and J. K. Percus, Phys. Rev., 155, 122 (1967).
    b) J. M. J. van Leeuwen and J. Weyland, Phys. Letters, 19, 562 (1966); Physica, 36, 457 (1967); 38, 35 (1968).
    c) W. R. Hoegy, Phys. Rev., 185, 210 (1969).
    d) E. H. Hauge and E. G. D. Cohen, Phys. Letters, 25A, 78 (1967); J. Math. Phys., 10, 397 (1969).

27. D. W. Jepsen, J. Math. Phys., 6, 405 (1965).
    G. Gallavotti, Phys. Rev., 185, 308 (1969).

28. See E. G. D. Cohen, in Lectures in Theoretical Physics, Vol. 8A, Univ. of Colorado Press, 1966, p.145; see also G. Sandri, R. O. Sullivan, and P. Harem, Phys. Rev. Letters, 13, 743 (1964) G. Sandri and A. H. Kritz, Phys. Rev., 150, 92 (1966).

29. J. S. Langer and T. Neal, Phys. Rev. Letters, 16, 984 (1966).

30. W. Wood and F. Lado, private communication.

31. J. Levine and T. M. Sanders, Phys. Rev., 154, 138 (1967).

32. H. E. Neustadter and M. H. Coopersmith, Phys. Rev. Letters, 23, 585 (1969).

33. Y. Pomeau, Phys. Letters, 27A, 601 (1968); Thesis, Univ. of Orsay, 1969.

34. B. J. Alder and T. E. Wainwright, Phys. Rev. Letters, 18, 988 (1969); J. Phys. Soc. Japan (Suppl.) 26, 267 (1968).

THE C^*- ALGEBRA APPROACH TO STATISTICAL MECHANICS

D. Ruelle

Institut des Hautes Etudes Scientifiques
Bures sur Yvette, France

During the past decade a large effort has been made to derive rigorous results in statistical mechanics. I want to describe here a special approach to this problem, the so-called C^*- algebra approach.

It is a fact that people who know something about statistical mechanics do not usually know much about C^*- algebras -- and vice-versa; this situation has led to a certain amount of skepticism on one side and to unjustified claims at the solution of big problems on the other side.

In this lecture I want to show that C^*- algebras provide a natural mathematical tool in the investigation of a class of conceptual problems arising in statistical mechanics. I shall be rewarded if I can convince Professor Uhlenbeck that the approach I am describing does contribute something to the understanding of the big problems of statistical mechanics.

I. STATES OF INFINITE CLASSICAL SYSTEMS

One of the main purposes of statistical mechanics is to describe the equilibrium states of a macroscopic system, knowing the microscopic interactions between its particles. The procedure to obtain these states is roughly the following: compute an equilibrium state for a finite system by one of the well-known ensembles, then let the system become infinite. Traditionally one would describe the state by its correlation functions (classical case) or reduced density matrices (quantum case).

Let us take a fresh look at the classical systems. What we mean by a state of a system is in fact a distribution of probability (also called a probability measure) on the possible configurations of the system. For an infinite classical system, a configuration ω is given by fixing the positions (and perhaps momenta) of all the particles in euclidean space. Let Ω be the space of the infinite configurations ω ; a state of an infinite system is then described by a probability measure ρ on Ω. Is it equivalent to give ρ and to give an infinite sequence of correlation functions? Yes, under suitable technical conditions. In any concrete computation, correlation functions will probably be superior, but the probability measure ρ is a conceptually simpler object.

Let A be a suitable function on Ω, we may write

$$< A > = \rho(A) = \int_{\Omega} \rho(d\omega)\, A(\omega)$$

and $<\cdot> = \rho(\cdot)$ is an expectation value in the sense that it is

linear: $\rho(\lambda A + \mu B) = \lambda \rho(A) + \mu \rho(B)$

positive: if $A \geq 0$ then $\rho(A) \geq 0$

normalized: $\rho(1) = 1$.

We have thus obtained a new way of describing a state: as an expectation value functional on an algebra $\mathcal{A}$ of functions on Ω. Notice that we may consider the function A as an observable of the system; the quantity $\rho(A)$ is then the expectation value of A in the state ρ.[1]

II. STATES OF INFINITE QUANTUM SYSTEMS

By analogy with the case of a classical system, a state ρ of a quantum system will be identified with an expectation value functional on an algebra $\mathcal{A}$ of "observables" of the system. In the present case the algebra $\mathcal{A}$ is no longer commutative so that ρ is no longer a probability measure on a space Ω of configurations of the system. What is $\mathfrak{A}$? It will consist of functions of the creation and annihilation operators $a(f)$, $a^*(f)$ for particles in the system (f is the wave function of a particle). It is mathematically convenient to take $\mathfrak{A}$ to be a C^*-algebra[2] i. e., an algebra over the complex numbers which satisfies a certain number of conditions, notably if a sequence of elements satisfies the Cauchy convergence criterion (with respect to norm) it has a limit on $\mathfrak{A}$. $\mathfrak{A}$ is called the C^*-algebra of the canonical commutation relations (for bosons) or anti-commutation relations (for fermions). An expectation value functional on a C^*-algebra is called a state on the C^*-algebra; we are thus describing a state of a physical system as a state on a C^*-algebra.

Let us take the expectation value, in the state ρ , of a product of creation and destruction operators:

$$D(x_1, \ldots, x_m ; y_1, \ldots, y_m)$$
$$= \rho(a^*(x_1) \ldots a^*(x_m) a(y_1) \ldots a(y_m)) \qquad (1)$$

What we get are just the reduced density matrices! What is then the advantage of talking about a state on a C^*-algebra? The advantage is technical and conceptual.

Technical: the expectation values in Eq. (1) need not make sense, because a, a^* are unbounded, and if they make sense, they may not determine completely the physical state ρ; with ρ defined on $\mathfrak{A}$ no such difficulty arises.

Conceptual: the description of a state as expectation value functional on a C*-algebra $\mathfrak{A}$ is not limited to the case of the canonical commutation and anticommutation relations, it works also for lattice spin systems, classical systems, etc.

One could ask: why not describe the state ρ by a vector or a density matrix in a Hilbert space as is usually done for quantum systems with a finite number of degrees of freedom? Here we meet an old and well-known difficulty, discussed in particular by Van Hove and Haag, and due to the existence of many inequivalent representations of the canonical (anti-) commutation relations.[3] If the Hilbert space $\mathfrak{H}$ is adequate for the description of a state ρ as vector or density matrix, it will in general not help for another state ρ'.[4] We emphasize that $\mathfrak{A}$ is an abstract C*-algebra, not considered as algebra of operators on any given Hilbert space. This corresponds to the fact that the canonical (anti-) commutation relations are algebraic equations, not involving any Hilbert space structure.

I will now give a few examples of the usefulness of the description of a physical state by a state on a C^*-algebra. I shall consider the following problems: time evolution of an infinite quantum system, decomposition of an equilibrium state into pure thermodynamic phases, and symmetry breakdown in statistical mechanics.

III. TIME EVOLUTION OF INFINITE QUANTUM SYSTEMS[5]

The time evolution of a system with a finite number of degrees of freedom is naturally investigated starting from the Schrödinger equation:

$$i\frac{\partial\psi}{\partial t} = H\psi \quad . \tag{2}$$

In the case of a system with an infinite number of degrees of freedom, this approach usually leads to insuperable difficulties because it is not known in what Hilbert space Eq. (2) should hold, or what H is.

Instead of deriving our inspiration from the Schrödinger picture, we shall get it from the Heisenberg picture by trying to find what the observable A will have become at time t; we may hope that this is again an element $A(t)$ of the C^*-algebra $\mathcal{A}$ This hope may be formulated more precisely: Suppose that for bounded space regions Λ a Hamiltonian H_Λ can be defined[6] such that, if $A \in \mathcal{A}$, we have again

$$A_\Lambda(t) = e^{iH_\Lambda t} A e^{-iH_\Lambda t} \in \mathcal{A}$$

suppose also that

$$A(t) = \lim_{\Lambda \to \infty} A_\Lambda(t)$$

exists, then $A(t)$ describes the time evolution of the observable A. This way of attacking the problem of time evolution does not always work; let me mention however that it permits the treatment, for instance, of the Heisenberg model of a ferromagnet.[7] An application of this result is the proof of an important property known as Kubo-Martin-Schwinger boundary condition: if $A, B \in \mathcal{A}$ and ρ is an equilibrium state corresponding to $\frac{1}{kT} = \beta$, there is a function F bounded and analytic in the strip $\{z : 0 < \operatorname{Im} z < \beta\}$ and with boundary values $F(t) = \rho(A(t)B)$, $F(t+i\beta) = \rho(BA(t))$ for t real. The state ρ can be obtained as a limit when $\Lambda \to \infty$ of the Gibbs equilibrium states for the finite regions Λ. I shall not discuss this limiting process but turn now to decompositions of an equilibrium state ρ, i.e., representations of ρ as averages of other states ρ_α:

$$\rho = \Sigma\, c_\alpha \rho_\alpha \quad \text{or} \quad \rho = \int d\mu(\alpha)\rho_\alpha \ .$$

IV. DECOMPOSITION OF AN INVARIANT EQUILIBRIUM STATE INTO PURE THERMODYNAMIC PHASES

I assume of course the translation invariance of the interactions for the physical systems considered. It is therefore natural to concentrate on states ρ which are translation invariant. For such a state, an entropy per unit volume $s(\rho)$ may be defined; this is the physical entropy if ρ is an equilibrium state. The fact that $s(\rho)$ depends on the state ρ only (it does not explicitly depend on the interaction) corresponds to the fact that the entropy is a measure of the amount of randomness present in ρ. The definition is in the quantum case

$$s(\rho) = \lim_{\Lambda \to \infty} \frac{-1}{V(\Lambda)} \operatorname{Tr} \rho_\Lambda \log \rho_\Lambda$$

where $V(\Lambda)$ is the volume of Λ and ρ_Λ is a density matrix describing the restriction of the state ρ to the bounded region Λ. The definition for the classical case is analogous.

One expects from thermodynamics that the (infinite volume) equilibrium states should make the following expression maximum:

$$s(\rho) + \beta\rho\,(\mu\mathcal{N} - \mathcal{E}_{\Phi}) \tag{3}$$

where the "observables" $\mathcal{N}$ and $\mathcal{E}_{\Phi}$ are respectively the number of particles and the energy with respect to the interaction Φ per unit volume. One can show that the infinite volume limits of Gibbs states satisfy this variational principle (this has been proved at least for lattice systems). More generally we shall say that the translation invariant states making Eq. (3) maximum are <u>invariant equilibrium states</u>.

The set Δ of invariant equilibrium states is what is called a <u>simplex</u>, i. e. , each point of Δ can be represented in a unique way as barycenter of extremal[8] points of Δ:

$$\rho = \sum_{\alpha} c_{\alpha}\rho_{\alpha} \quad \text{or} \quad \rho = \int d\alpha\, \rho_{\alpha} .$$

This decomposition can be interpreted as decomposition of the "mixture" ρ into "pure thermodynamic phases" ρ_{α}. This interpretation is confirmed by the fact that in a pure phase thus defined the macroscopic observables (like density) have very small fluctuations, while the fluctuations are large for true mixtures.

V. FURTHER DECOMPOSITION. THE PROBLEM OF SYMMETRY BREAKDOWN

The decomposition just discussed is into states ρ_α which are again invariant equilibrium states. It is interesting to try to decompose the state ρ into equilibrium states that need not be translation invariant any more. For instance, one expects that if an invariant equilibrium state ρ describes a crystal, it can be decomposed into states ρ_α where the orientation and mean positions of the atoms are fixed (ρ_α is now periodic rather than invariant under translations).

How does one define a (generally noninvariant) equilibrium state? For classical systems it appears that the answer can be given in terms of equilibrium equations.[9] Let Λ be a bounded space region. A state ρ yields a probability $P_\Lambda(X, Y)$ that the infinite system is in a configuration X inside Λ and Y outside Λ. In particular $P_\Lambda(\emptyset, Y)$ is the probability that there are no particles in Λ and that a configuration Y occurs outside of Λ. If ρ is an equilibrium state for the interaction Φ one expects that

$$P_{\Lambda}(X, Y) = \mathcal{z}^{N(X)} e^{-\beta U_{\Phi}(X) - \beta W_{\Phi}(X, Y)} P_{\Lambda}(\emptyset, Y) \qquad (4)$$

where $\mathcal{z}$ is the activity, $N(X)$ and $U_{\Phi}(X)$ are the number of particles, and the energy of the configuration X, $W_{\Phi}(X, Y)$ is the extra energy due to the interaction of X and Y. The Eqs.(4) are what I call the equilibrium equations.

Having introduced general equilibrium states, we may define symmetry breakdown. A pure thermodynamic phase is said to exhibit symmetry breakdown if it has a nontrivial decomposition into noninvariant equilibrium states. As indicated earlier this is expected to happen for a state describing a crystal. For some lattice systems, symmetry breakdown has been rigorously proved to occur by Dobrushin. His nicest example corresponds to particles with nearest neighbor exclusion where at high density the invariant equilibrium state is a mixture of two periodic (but noninvariant) equilibrium states.

The equilibrium equations have no straightforward generalization to quantum systems, but evidence is accumulating to indicate that the states satisfying the KMS condition are those one should define as equilibrium states.[10]

REFERENCES

1. A more traditional way of speaking is to call A a "random variable" rather than "observable."

2. The use of C*-algebras was put forward by Segal,[17] Haag and Kastler[13] and others.

3. In the case of infinitely many degrees of freedom.

4. In particular by making a perturbation on a system, one usually forces the system out of its original Hilbert space. Part of the infinities of quantum field theory comes from there.

5. A more detailed discussion of the problems in this and in the next section is given in Ruelle.[16]

6. We suppose that $H \in \mathfrak{A}$, but less restrictive assumptions are possible.

7. Robinson.[15]

8. An extremal point of Δ is a point which is not the middle point of any open line segment contained in Δ. The theory of infinite-dimensional simplexes is due to G. Choquet.

9. See Dobrushin,[11] and Lanford and Ruelle.[14]

10. KMS states are discussed by Haag, Hugenholtz, and Winnink.[12] They turned out recently to be extremely interesting from a purely mathematical viewpoint (see Takesaki[18]).

11. R. L. Dobrushin, Teorija Verojatn. i ee Prim., 13, 201 (1968); Funkts. Analiz i ego Pril., 2, 31 (1968); 2, 44 (1968); 3, 27 (1969).

12. R. Haag, N. Hugenholtz, and M. Winnink, Commun. Math. Phys., 6, 215 (1967).

13. R. Haag and D. Kastler, J. Math. Phys., 5, 848 (1964).

14. O. E. Lanford and D. Ruelle, Commun. Math. Phys., 13, 194 (1969).

15. D. W. Robinson, Commun. Math. Phys., 6, 151 (1967); 7, 337 (1968).

16. D. Ruelle, Statistical Mechanics, Benjamin, New York, 1969.

17. I. Segal, Mathematical Problems of Relativistic Physics, Amer. Math. Soc., Providence, 1963.

18. M. Takesaki, Tomita's theory of modular Hilbert algebras and its applications, Springer, Berlin, 1970.

THE CURIE POINT

C. Domb

King's College University of London
London, England

I. INTRODUCTION

I am particularly happy to be associated with this symposium in honor of Professor Uhlenbeck for at least three reasons:

a) It is not generally known how much his ideas have influenced detailed recent developments in exploring the Curie point.

b) In his introduction to the Washington Conference on "Critical Phenomena" in 1965, Uhlenbeck[1] told us something of the historical background to the subject and stimulated me to investigate a little further. I shall report a few of the results in a moment.

c) In the same talk Uhlenbeck suggested to us that one of

the great challenges today was "the reconciliation of Onsager with Van der Waals." In the four and a half years that have elapsed since the Washington Conference, we have gone some way toward effecting such a reconciliation, and this symposium is surely an appropriate place for a review.

In view of the title assigned to me, I thought that I ought at least to read Curie's famous 1895 paper[2] which, I assumed, initiated the subject of my talk. To my great surprise, although it contained very much of interest and importance, there was nothing which could be positively identified with the discovery of the Curie point. In fact, Curie says that the expression "Transformation temperature of Iron" which had been commonly used is not a precisely defined temperature at all but should be replaced by the average of the temperatures corresponding to points of inflection of the magnetization curves.

Curie's statement was quite correct, since he was considering the behavior in a magnetic field. Elsewhere he states more specifically:

> "The magnetic transformation near 750° is not sharp, at least for intense fields. The diminution in intensity of magnetization is only extremely rapid, becoming increasingly rapid, as the field becomes

> increasingly weak. It tends to a sharp transformation when the fields become extremely weak."

Curie refers back to experiments by Hopkinson, and Uhlenbeck had himself stated, "So far as I know it was only in 1890 that Hopkinson found that iron above a certain temperature ceased to be ferromagnetic." I therefore looked back to Hopkinson's papers of which there are a considerable number and found them of great interest. Clearly he was the first to use the term "critical temperature" in connection with a ferromagnet; in his Royal Society paper[3] in 1889, Hopkinson refers to "the temperature at which the magnetism disappears which we may appropriately call the critical temperature."

Incidentally, perhaps I may be allowed a small digression on the subject of Hopkinson.[4] He was a very gifted mathematician and experimenter who was elected to the Royal Society at the age of 29 and received a Royal Medal in 1890 at the age of 40. Unfortunately he died tragically with three of his children in a climbing accident in 1898. For many years he was Professor of Electrical Engineering at my own college, King's College, London, and therefore, I have a certain personal interest in his reinstatement in the literature.

But although Hopkinson was the first to use the term "critical temperature," it is clear from his paper that the phenomenon of the disappearance of magnetization at high temperatures was known long before. Hopkinson referred back to Bauer[5] in 1880 and Barrett in 1874, who in turn referred back to Faraday in the 1830's, and the chain led eventually to Gilbert who, in his famous treatise[6] "De Magnete" in 1600 stated that, "Magnets lose their strength in a furnace."

How did the term "Curie point" arise? An important feature which emerged from Curie's investigations was the law that the susceptibility of a paramagnet is inversely proportional to the absolute temperature $\chi_o = C/T$, and hence this was referred to as Curie's law and the constant C as Curie's constant. In 1907 Pierre Weiss[7] published his famous theoretical paper on ferromagnetism in which he introduced his "molecular field hypothesis" and this established the existence of a mathematical point at which the spontaneous magnetization in zero field disappeared. Instead of Curie's law, Weiss suggested that the susceptibility of a ferromagnet above this termperature satisfied the relation $\chi_o = C/(T-\theta)$. In the years which followed, Pierre Weiss launched a detailed experimental program to test his theory, and in a joint paper[8] with Kamerlingh-Onnes in the Journal de Physique in 1910 the term "Point de Curie" first appears with the following footnote:

> "C'est ainsi que nous appelerons dans ce travail la température de perte du ferromagnétisme spontané. Il n'y a là aucune contradiction avec l'idée de température de transformation fonction du champ enoncée par Curie, la température de perte du ferromagnétisme spontané etant celle que l'on détermine en faisant tendre le champ vers zéro."

Weiss continued to use this terminology in all his subsequent papers and in his book with Foex Le Magnétisme published in 1926.

However, the term was not taken up in England for many years. Here are some typical quotations:

a) Encyclopaedia Britannica, 11th edition, 1911; Shelford Bidwell:

> "It has long been known that iron when raised to a certain "critical temperature" corresponding to dull red heat, loses its susceptibility and becomes magnetically indifferent, or, more accurately, is transformed from a ferromagnetic to a paramagnetic body. Recent researches have shown that other important changes in its properties occur at the same critical temperature. Abrupt alterations take place in its density, specific heat, thermo-electric quality, electrical conductivity, temperature-coefficient of electrical resistance, and in some at least of its mechanical properties. Ordinary magnetizable iron is in many respects an essentially different substance from the non-magnetizable metal into which it is

transformed when its temperature is raised above a certain point (see Brit. Assoc. Report 1890; 145)."

b) Dictionary of Applied Physics, Glazebrook 1922, Vol. II, page 512-3, article on "Modern Theories of Magnetism" by Arthur Ernest Oxley:

> "For each ferromagnetic substance there is a certain temperature above which this mutual cooperation ceases to have these remarkable magnetic effects, the ferromagnetic substance having changed into a paramagnetic substance, and following more or less closely the Curie rule for paramagnetic substances. This temperature is known as the critical temperature."

c) Encyclopaedia Britannica, 13th edition, 1926, article by O.W. Richardson:

> "This gives a single intensity of permanent magnetization which is stable, but which disappears, and with it the ferromagnetism at a definite temperature, the critical temperature."

It is only in the 14th edition in 1929 that the article by E.C. Stoner makes complete use of the "Curie point" terminology (nevertheless, he uses the term "Weiss' Law" for $\chi_o(T-\theta) = C$ which was also the terminology used by O.W. Richardson).

By 1934, the terminology was sufficiently widespread for Bragg and Williams to employ it in their analogous theory for order-disorder transitions.[9]

Without underrating the great significance of Curie's classical paper, the influence of which I shall discuss in the next section, I think the terminology is overgenerous to him and unfair to his predecessors like Hopkinson who knew of the existence of a critical temperature long before. What can we do to restore the balance? We should at least see that Hopkinson receives his proper place in the literature, and Weiss his own share of the credit by associating the appropriate law with his name and calling θ in the above equation the "Weiss temperature." However, I think that the term "Curie point" is by now too well established for any change to be contemplated.

II. HISTORICAL SURVEY

Of the significance of Curie's 1895 paper, Stoner[10] writes as follows:

> "All the observations had been in a sense isolated. A comprehensive experimental survey of a wide range of substances under widely varied conditions was needed before much further progress could be made in elucidating the signficance of magnetic phenomena. It is such a survey which was provided by Curie's work, which is worthy to rank among the great classical experimental researches .... The whole investigation may be said to form the experimental foundation for modern theoretical work on magnetism."

From our point of view, one of the most interesting features of the paper is the final section which draws attention to the analogy between fluids and magnets. Taking pressure p, as the analogue of magnetic field H, and density ρ, as the analogue of magnetization M, Curie points out the close similarity between the $p-\rho$ and $M-H$ isothermals. The paramagnetic state at high temperatures corresponds to the gaseous phase and the ferromagnetic state at low temperatures to the liquid phase. Curie says that this analogy could be used to suggest new and useful experiments, and he poses the question whether there exists a precisely defined critical point with associated critical constants for a ferromagnet analogous to a fluid. It was this analogy which led Pierre Weiss in 1907 to postulate his "Molecular Field" hypothesis[7] in which he assumed that the mutual interactions between molecules can be replaced by a uniform field nM, proportional to the magnetization and in the same direction. He states, "One may give to nM the name "Internal Field" to mark the analogy with the internal pressure of van der Waals." We thus see clearly that his aim was to produce a van der Waals theory for magnetism. The well known theory which he developed was outstandingly successful and accounted satisfactorily for all of Curie's experimental observations. However, the analogy between magnets and fluids seems to have been forgotten for more than 30 years until it was rediscovered in the first lattice gas model of Cernuschi and Eyring[11] in 1939. In recent years it has been one of the most fertile ideas for exploring detailed behavior in the critical region.

Let me briefly mention in passing the theory of order-disorder transitions in alloys, particularly as formulated by Bragg and Williams[9] in 1934. As mentioned above, they used the term "Curie point" to describe the transition temperature, although the true analogue of an alloy is an antiferromagnet. It is true that there are similarities between this phenomenon and a ferromagnet, but there are also important differences, and I shall therefore not discuss antiferromagnetism in the present survey.

The next major advance occurred in 1937-9 with the development of the cluster integral theory of condensation. The first important step had been taken by Ursell[12] in 1927, but the most significant progress was due to Mayer[13] and his collaborators. (However, a parallel development by Yvon[13a] overlapped considerably.)

The cluster integral theory was undoubtedly a landmark in statistical mechanics. It provided a closed form expression for *all* the virial coefficients, and when it was developing there was a general hope among theoreticians that it would be able to account for all the features on condensation and critical behavior. Uhlenbeck has commented[1] that in 1938 at an International Congress to commemmorate van der Waals' birthday his equation was mentioned by only one of the participants; all of the others were enthusiastically concerned with the new so-called "rigorous" development. It was in the 1950's that Uhlenbeck[13b] and his collaborators demonstrated clearly that the above hope was illusory, and this will be discussed in more detail in the next section.

It is fitting to terminate this survey with the publication which inaugurated a new era in the theory of critical phenomena, Onsager's famous solution in 1944 of the two dimensional Ising model.[14] This model had been introduced by Ising[15] on the recommendation of his director of studies, Lenz, in 1926, but Ising had been able to get no further than one dimension. In two dimensions, Peierls[16] had shown in 1936 that there was a Curie point at a finite temperature, and

Kramers and Wannier[17] had located the Curie point exactly in 1941. Onsager calculated all the thermodynamic properties of the model in zero field, and subsequently with Kaufmann[18] he obtained important information on the behavior of the correlations. A calculation of the long-range order (spontaneous magnetization) followed soon after.[19] It would be difficult to exaggerate the importance of this paper.

Nevertheless, one must also point out what the paper did not give by camparison with the earlier van der Waals or Weiss theories. It did not provide even for the two dimensional Ising model a complete equation of state. But it did provide a rigorous calculation of a selected number of thermodynamic properties which showed clearly that near the Curie point all classical calculations were wrong.

Since Onsager's paper, there have been no major landmarks of comparable significance, but a process of erosion has taken place. Problems have been worn down rather than solved, and in the past decade this erosion has taken place at a rapidly increasing rate. By now we have assembled enough information to suggest a reconciliation between Onsager and van der Waals in response to Uhlenbeck's challenge.

III. LATTICE MODELS

Most of the recent progress in the exploration of the Curie point has resulted from an abandonment of the continuum model for a lattice model. A large variety of lattice models have been considered using Ising and Heisenberg interactions of varying spin values with short- or long-range forces in two, three, and even higher dimensions. The latter may have no direct physical application, but they help us to understand the nature of the problem we are tackling and how its solution depends in a general way on dimension.

Most theoreticians are apologetic about using a lattice model for a fluid. However, following Kirkwood,[20] the model can be justified rigorously if we assume a hard core potential. We can then always impose an artificial lattice mesh of such a size that each cell can contain only zero or one particle. The interaction between cells can take on a continuum of values, and there will be no symmetry in a magnetic field of the type associated with the Ising and Heisenberg models. However, if we study the properties of lattice models for a variety of interactions, we may be able to select one which provides a reasonable approximation to a fluid.

Why has so much progress been achieved with lattice models, whereas the cluster integral theory provided no detailed information regarding critical behavior? To answer this question, let me refer in more detail to the researches of Uhlenbeck and his collaborators[21] in the 1950's, the general message of which was "if graphs play a major part in theory, one must treat them seriously and study them in their own right." Hence they provided us with a systematic enumeration of connected graphs and stars, and tried to obtain information regarding the asymptotic behavior of the numbers of these graphs. The key to the whole development lay in Polya's theorem,[22] and for labeled graphs and stars they were able to obtain asymptotic formulae which showed that the elimination of nonstar graphs in the Mayer theory gave little numerical advantage. In regard to unlabeled graphs, they were able to deal in a general manner only with connected graphs; the enumeration of unlabeled stars remained for many years an unsolved problem. (A solution has recently been obtained by Robert W. Robinson.[23]) But they showed that the number of stars increased with extraordinary rapidity with the number of points p, and this phenomenon is shown in Table 1. In fact it is easy to obtain a lower limit for S(p) the number of unlabeled stars of p points by taking the asymptotic formula for the number of labeled graphs and dividing by p! which is the maximum symmetry factor. (This should be a good approximation since Uhlenbeck and Ford[21] showed that for large p nearly all graphs have no symmetry.) The result of this estimate is shown in the last row of Table 1.

When we remember that for each star we must evaluate a multi-dimensional integral, it is not surprising that even

TABLE 1

NUMBER OF GRAPHS WITH p POINTS

	C(p) Connected graphs					S(p) Stars		
p	3	4	5	6	7	8	9	10
C(p)	2	6	21	112	853	11117	261080	
S(p)	1	3	10	56	468	7123	194066	
$\frac{2^{\frac{1}{2}p(p-1)}}{p!}$	1.3	2.7	8.5	45.5	416.2	6658	189372	9695870

S(8) and S(9) have been taken from Robert W. Robinson.

with the best computers available not many terms of the virial expansion have been evaluated. Rushbrooke[24] has pointed out that for the simplest hard sphere potential all the exact results were already obtained by Boltzmann in 1899. Approximation and Monte Carlo methods have allowed us to add a total of three new terms. I am prepared to take the risk of predicting that within the next decade no more than one new term will be added, and possibly not even one.

The specific gains of a lattice model can be listed as follows:

a) Instead of topological graphs we consider graphs embedded in the lattice constructed from bonds which correspond to interactions; instead of evaluating cluster integrals, we have to count the number of possible embeddings, and computers can be used in a very positive way.

b) Many of the stars correspond to zero embeddings, particularly for lattices with low coordination like the simple quadratic and diamond. Hence, one can contemplate the possibility of 15 or 20 terms of a Mayer series.

c) For lattice models the most important results have been obtained from high temperature expansions of the partition function of the form

$$Z_N = \langle \exp -\beta \mathcal{H} \rangle = 1 - \beta \langle \mathcal{H} \rangle + \frac{\beta^2}{2!} \langle \mathcal{H}^2 \rangle \ \ldots \frac{(-\beta)^\ell}{\ell!} \langle \mathcal{H}^\ell \rangle \ldots \quad (1)$$

The grouping of the graphs is according to lines ℓ rather than points p, and Table 2 shows that (a) the number of connected graphs $C(\ell)$ rises much more slowly and (b) there is a much greater gain in eliminating nonstar graphs.

d) The symmetry of the partition function to reversal of magnetic field gives rise to many simplifying features.

e) We can look at many different lattices with coordination numbers varying from 3 for the hydrogen peroxide lattice[25] to 12 for the face centered cubic lattice. For a given property convergence is usually more rapid for a particular lattice. In any case we can search for "long-range" properties which are lattice independent.

TABLE 2

NUMBER OF GRAPHS WITH ℓ LINES

$C(\ell)$ Connected graphs $S(\ell)$ Stars

ℓ	3	4	5	6	7	8	9	10	11	12
$S(\ell)$	1	1	2	4	7	16	42	111	331	1094
$C(\ell)$	3	5	12	30	79	227	710	2322		

Following Uhlenbeck's lead, my colleague, Dr. M. F. Sykes,[26] has suggested the classification of star graphs by means of a cyclomatic number $c = (\ell - p + 1)$ which specifies the general shape or topology of the graph. The only topology of cyclomatic number 1 is a simple polygon; there is also only one topology of cyclomatic number 2 consisting of a polygon with a bridge which Sykes has termed a Theta-graph; there are four topologies of cyclomatic number 3, and 17 of cyclomatic number 4 which are illustrated in Fig. 1. Now, for example, if we wish to enumerate embeddings or "lattice constants" of all graphs of type θ with 12 lines, we need only list all elementary partitions of 12 into 3 numbers (r, s, t). For higher order graphs we need to combine partitions with symmetry, but the procedure can be standardized.

In order to pursue this classification to values of $c > 4$, Heap[27] has shown how we can make use of computers. There are 118 topologies with $c = 5$ and 1198 with $c = 6$, and Heap has pointed out that there is no great benefit (except of personal gratification) in drawing diagrams of all the graphs. It is far better to store them in the computer since this eliminates a possible source of error.

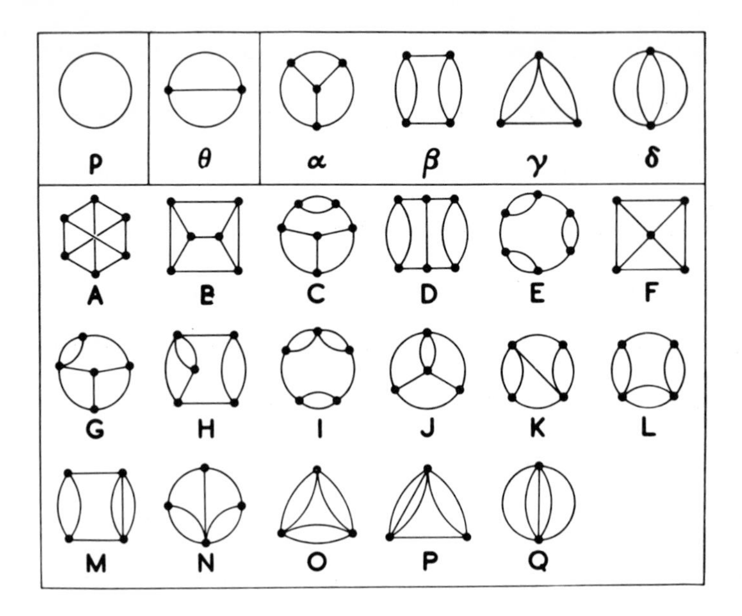

FIG. 1

Graph topologies with $c = 1\text{-}4$ (following M. F. Sykes).

Let me digress here to mention another classification of these topologies which has proved useful at least conceptually.[28] Consider first a lattice of coordination number 3. The maximum number of lines which can meet at a lattice point is 3, and the only topologies in Fig. 1 which are relevant to such a lattice are p; θ; α, β; and A, B, C, D, E. All of these topologies can be described as a polygon with $(c-1)$ bridges. For example, with $c = 4$, if we take 6 points (1, 2, 3, 4, 5, 6) on a circle and connect them up in pairs in all possible ways, we obtain $6!/3!(2!)^3 = 15$ possibilities. But of these 15 there are only 5 distinct topologies corresponding to A, B, C, D, E.. If we now allow points of the different pairs to coalesce, we run through all the remaining 12 topologies. However, this method of description breaks down at $c = 5$; there are now 16 topologies with nodes of order 3, but one of them (numbered 105 in Fig. 2) cannot be described as a polygon with bridges; when nodes of higher order are considered, 5 additional topologies do not fit the description (Fig. 2).

In Uhlenbeck's description of the graph theoretical background of the Mayer theory, he focussed attention on two theorems. By means of the first of these, all disjoint graphs could be eliminated, and the theorem applied widely to many body forces of interaction and to bulk and surface properties. The second theorem which eliminated nonstar graphs applied in more restricted conditions. Both theorems have their analogues in high temperature expansions. The first, which eliminates disjoint embeddings, applies very widely and corresponds in statistical terminology to the change from a moment to a cumulant expansion

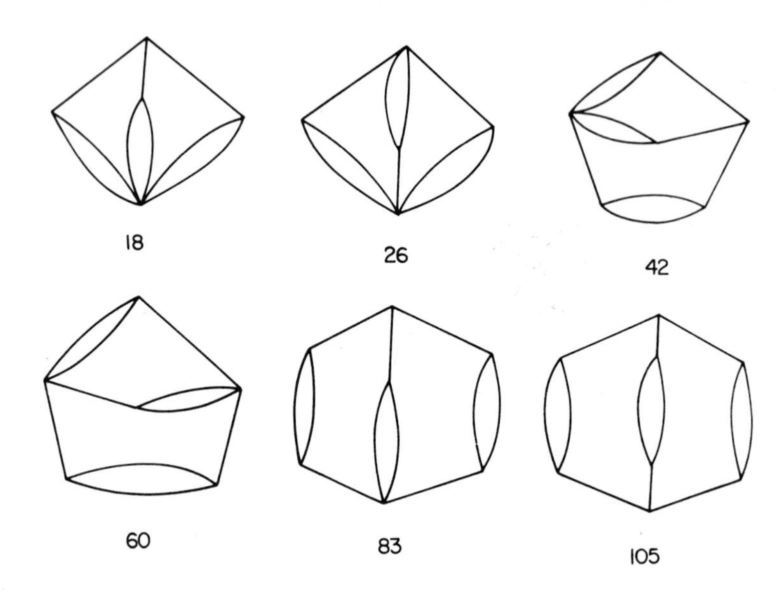

FIG. 2

Graph topologies with $c = 5$ which cannot be represented as bridged polygons. The numbers follow Heap's list. (Ref. 27).

$$Z_N = \sum_{\ell, x} [G_{\ell x}] g_{\ell x}(\beta J, mH) \tag{2}$$

$$\ln Z_N = \sum_{\ell, x} [C_{\ell x}] c_{\ell x}(\beta J, mH) \ . \tag{3}$$

Here $G_{\ell x}$ represents <u>all</u> graphs of ℓ lines and $C_{\ell x}$ all <u>connected</u> graphs of ℓ lines (x being a labeling suffix). The symbol [ ] is used for the number of embeddings; and $g_{\ell x}$, $c_{\ell x}$ are weight functions depending on the particular model. The second which eliminates nonstar embeddings is restricted to the Ising, planar classical Heisenberg, classical Heisenberg, and in general to rotating classical vectors in r dimensions, in zero field.

$$\ln Z_N = \sum_{\ell, x} [S_{\ell x}] s_{\ell x}(\beta J) \quad . \tag{4}$$

Here $S_{\ell x}$ represents all star graphs of ℓ lines, and $s_{\ell x}$ is a weight function.

From equations (2) and (3) it will be seen that it is necessary to construct tables of embeddings of "lattice constants" which can then be used for a variety of different models. If we need consider only star graphs, the tables can be extended further. An example of such a tabulation by our own group at King's College is given in Table 3. Lattice constants for individual graphs were evaluated by using a computer programme due to J. L. Martin. When all connected graphs need to be considered, the numbers increase much more rapidly and it is only possible to reach smaller values of ℓ. Comprehensive tables of lattice constants for connected graphs of up to 10 lines have been

TABLE 3

LATTICE CONSTANST OF β GRAPHS OF ℓ LINES

Beta Graphs		T	FCC
ℓ = 8	1. 2. 1. 2. 1. 1	6	564
ℓ = 9	1. 2. 1. 2. 1. 2	30	6 696
	1. 2. 1. 3. 1. 1	24	5 472
	1. 2. 2. 2. 1. 1	0	1 656
ℓ = 10	1. 2. 1. 2. 1. 3	102	46 128
	1. 2. 1. 2. 2. 2	51	23 052
	1. 2. 1. 3. 1. 2	120	64 560
	1. 2. 1. 4. 1. 1	60	32 760
	1. 2. 2. 2. 1. 2	12	21 264
	1. 2. 2. 3. 1. 1	6	16 992
	1. 3. 1. 3. 1. 1	24	13 470
	1. 3. 2. 2. 1. 1	0	7 968
	2. 2. 2. 2. 1. 1	0	1 224
ℓ = 11	1. 2. 1. 2. 1. 4	330	345 192
	1. 2. 1. 2. 2. 3	330	346 392
	1. 2. 1. 2. 1. 3	360	436 224
	1. 2. 1. 3. 2. 2	180	219 312
	1. 2. 1. 4. 1. 2	276	380 784
	1. 2. 1. 5. 1. 1	168	219 216
	1. 2. 2. 2. 1. 3	60	159 264
	1. 2. 2. 2. 2. 2	30	79 992
	1. 2. 2. 3. 1. 2	60	224 016
	1. 2. 2. 4. 1. 1	30	110 208
	1. 2. 3. 3. 1. 1	0	41 376
	1. 3. 1. 3. 1. 2	120	155 256
	1. 3. 1. 4. 1. 1	120	160 320
	1. 3. 2. 2. 1. 2	24	101 808
	1. 3. 2. 3. 1. 1	12	81 576
	1. 4. 2. 2. 1. 1	0	47 256
	2. 2. 2. 2. 1. 2	0	17 616
	2. 2. 2. 3. 1. 1	0	25 152
ℓ = 12	1. 2. 1. 2. 1. 5	1 164	2 707 560
	1. 2. 1. 2. 2. 4	1 170	2 718 480

Key: a (b — e / f — c) d = (a, b, c, d, e, f) .

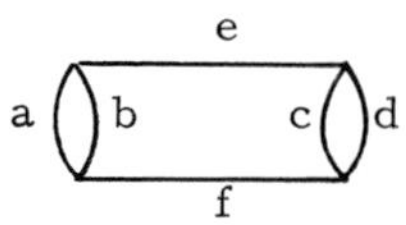

FOR THE TRIANGULAR AND FCC LATTICES

Beta Graphs	T	FCC
1. 2. 1. 2. 3. 3	588	1 362 060
1. 2. 1. 3. 1. 4	1 104	3 201 888
1. 2. 1. 3. 2. 3	1 104	3 232 176
1. 2. 1. 4. 1. 3	816	2 521 968
1. 2. 1. 4. 2. 2	408	1 271 856
1. 2. 1. 5. 1. 2	756	2 511 408
1. 2. 1. 6. 1. 1	516	1 587 288
1. 2. 2. 2. 1. 4	288	1 258 848
1. 2. 2. 2. 2. 3	300	1 269 312
1. 2. 2. 3. 1. 3	276	1 671 120
1. 2. 2. 3. 2. 2	138	841 392
1. 2. 2. 4. 1. 2	264	1 452 576
1. 2. 2. 5. 1. 1	126	802 128
1. 2. 3. 3. 1. 2	0	567 648
1. 2. 3. 4. 1. 1	0	538 416
1. 3. 1. 3. 1. 3	312	1 025 220
1. 3. 1. 3. 2. 2	156	520 554
1. 3. 1. 4. 1. 2	552	1 820 544
1. 3. 1. 5. 1. 1	336	1 067 364
1. 3. 2. 2. 1. 3	120	744 336
1. 3. 2. 2. 2. 2	60	376 872
1. 3. 2. 3. 1. 2	120	1 060 944
1. 3. 2. 4. 1. 1	60	523 536
1. 3. 3. 3. 1. 1	0	196 536
1. 4. 1. 4. 1. 1	150	474 180
1. 4. 2. 2. 1. 2	60	593 520
1. 4. 2. 3. 1. 1	30	478 656
1. 5. 2. 2. 1. 1	0	312 816
2. 2. 2. 2. 1. 3	12	141 948
2. 2. 2. 2. 2. 2	9	72 786
2. 2. 2. 3. 1. 2	24	372 672
2. 2. 2. 4. 1. 1	6	163 440
2. 2. 3. 3. 1. 1	0	61 200
2. 3. 2. 3. 1. 1	6	129 360

provided in a Brookhaven Report by Baker and collaborators.[29]

We terminate this section by quoting three typical series expansions for the magnetic susceptibility which have been obtained by the above methods.

1. Triangular Lattice Susceptibility. Ising Model. Spin 1/2

$$\frac{\chi(v)}{N\beta m^2} = 1 + 6v + 30v^2 + 138v^3 + 606v^4 + 2586v^5 + 10\,818v^6 + 44\,574v^7 + 181\,542v^8 + 732\,678v^9 + 2\,935\,218v^{10} + 11\,687\,202v^{11} + 46\,296\,210v^{12} + 182\,588\,850v^{13} + 717\,395\,262v^{14} + 2\,809\,372\,302v^{15} + 10\,969\,820\,358v^{16} + \ldots \quad (v = \tanh \beta J) \tag{5}$$

2. Simple Cubic Susceptibility. Ising Model. Spin 1/2

$$\frac{\chi(v)}{N\beta m^2} = 1 + 6v + 30v^2 + 150v^3 + 726v^4 + 3\,510v^5 + 16\,710v^6 + 79\,494v^7 + 375\,174v^8 + 1\,759\,686v^9 + 8\,306\,862v^{10} + 38\,975\,286v^{11} + 182\,265\,822v^{12} + 852\,063\,558v^{13} + 3\,973\,784\,886v^{14} + 18\,527\,532\,310v^{15} + 86\,228\,667\,894v^{16} + 401\,225\,391\,222v^{17} + \ldots \quad (v = \tanh \beta J) \tag{6}$$

3. FCC Susceptibility. Heisenberg Model. Spin 1/2

$$\frac{\chi(o)}{N\beta m^2} = 1 + 12K + 240\,K^2/2! + 6\,624\,K^3/3! + 234\,720\,K^4/4! + 10\,208\,832\,K^5/5! + 526\,810\,176\,K^6/6! + 31\,434\,585\,600\,K^7/7! + 2\,127\,785\,025\,024\,K^8/8! + 161\,064\,469\,168\,128\,K^9/9! \; 161\,064\,469\,168\,128\,K^9/9! + \ldots \quad (K = \beta J) \tag{7}$$

Data for Eqs. (5) and (6) have been supplied by M. F. Sykes; equation (7) is taken from a paper by Baker, Gilbert, Eve and Rushbrooke.[30]

(Underscored numbers represent terms which have been added since the original estimation of the critical exponents.[30a])

IV. CRITICAL BEHAVIOR

Nearly all high temperature series of the type illustrated in Eqs. (5), (6), and (7) are smooth and well behaved, particularly for lattices of high coordination. Low temperature series often contain spurious singularities and are more difficult to deal with. I shall not enter into a detailed exposition of methods which have been used to conjecture critical behavior from such expansions, since I discussed this at length in my contribution to the Washington Conference.[31] However, I should like to raise a point of general principle.

I have discussed the use of numerical data of this type with professional statisticians who assure me that no meaningful analysis can be undertaken without some preconceived notion of the form of the function we are looking for. In fact, we do start with such a notion which we obtain by making use of a small number of rigorous results, together with "physical intuition." For example, we are prepared from our knowledge of the model and experimental results associated with it to suggest that the susceptibility is singular at the same temperature as the specific heat. Once we have adopted such a hypothesis, we can use the numerical data to test its validity. Or we might be prepared to try the hypothesis that the analytic character of the partition function is the same in three dimensions as in two dimensions. As more terms become available, such hypotheses can be subjected to more and more stringent tests.

In this manner a collection of values of "critical exponents" for various models was gradually assembled,[32] which could be compared with experiment. Thus, the series Eqs. (5), (6), and (7) could be well fitted by the form

$$\chi_o \sim A(1 - T_c/T)^{-\gamma} \tag{8}$$

with $\gamma = 7/4$ for (5), $\gamma = 5/4$ for (6), and $\gamma \cong 1.43$ for (7). The most important general conclusion emerging theoretically was that although exponents like γ varied from model to model, they did not vary with lattice structure in a given dimension. However, they usually differed from their classical counterparts. Exponents corresponding to different properties (e.g., specific heat, low temperature susceptibility) were all treated in isolation of one another.

The important suggestion that there might be relations between critical exponents came from an investigation of the droplet model by Essam and Fisher.[33] Although this model ignores the "volume exclusion" of different droplets, and therefore has no validity near the Curie point, it nevertheless opened a very important door and led the way to the postulation of a general equation of state which could reconcile Onsager with van der Waals.

This equation of state was suggested simultaneously by three different groups using independent approaches; Widom[34] put forward a "homogeneous" equation, and Kadanoff[35] provided a justification in the form of a "scaling" hypothesis; Patashinskii and Pokrovskii[36] investigated the form of the multiple correlation function in high temperature expansions and used this to conjecture the M-H relation in nonzero field; Hunter and I[37] investigated the higher derivatives of lnZ with respect to magnetic field as $T \to T_{c+}$ and found that there seemed to be a single "gap" index between successive derivatives; if such an index was assumed an M-H equation of state could easily be derived. The most clear and elegant formulation of this equation of state is due to Griffiths,[38]

$$H = M^{\delta} h(tM^{-1/\beta}) \quad [t = (T/T_c - 1)] \ . \tag{9}$$

Critical behavior for any model is thus determined by two parameters β, δ (in terms of which all other critical exponents can be calculated) and a single function $h(x)$ assumed analytic. $h(x)$ is defined only in the range $-x_0 < x < \infty$. It has been conjectured[39] that β, δ, and $h(x)$ depend only on dimension and model but not on lattice structure.

For the classical theories of Weiss and van der Waals $\beta = \frac{1}{2}$, $\delta = 3$, and $h(x)$ takes on a particularly simple linear form

$$h(x) = 1 + ax \ . \tag{10}$$

These theories have now been shown to be rigorously correct for a system with very long range forces.[40] The possibility of varying β, δ, and $h(x)$ enables us to describe the critical behavior of the various models and interactions introduced at the beginning of the previous section.

We can obtain a useful qualitative picture of how β, δ, and $h(x)$ vary with range of force from the spherical model. This model was introduced for short-range forces by Berlin and Kac[41] in 1952, and consisted of Ising spins which could take any values subject to the restriction

$$s_1^2 + s_2^2 + \ldots s_n^2 = N \quad . \tag{11}$$

Although Berlin and Kac were able to derive an exact solution for the model, it was difficult to find a suitable physical interpretation. In 1966 Joyce[42] showed how the solution could be extended to long-range forces of the form

$$J_{ij} = J/R_{ij}^{d+\sigma} \tag{12}$$

in dimension d; by varying d and σ a considerable variation of critical exponents could be achieved, and when the forces became sufficiently long range, the classical form was retrieved.

The significance of the spherical model became clearer when Stanley[43] pointed out that it corresponded to a classical vector in r dimensions as $r \to \infty$.

Using the method introduced by Joyce it is possible to determine[44] the critical behavior for all forces given by Eq. (12) and to show that Eq. (9) is satisfied exactly with the following values:

$$\beta = \frac{1}{2}\,; \qquad h(x) = (1 + ax)^{\gamma}, \qquad \delta = 1 + \gamma/\beta \tag{13}$$

$$d = 1\,:\quad \gamma = 1 \quad (0 < \sigma < \tfrac{1}{2});\quad \gamma = \sigma/(1-\sigma)(\tfrac{1}{2} < \sigma < 1); \tag{14}$$

$$d = 2\,:\quad \gamma = 1 \quad (0 < \sigma < 1);\quad \gamma = \sigma/(2-\sigma)(1 < \sigma < 2); \tag{15}$$

$$d = 3\,:\quad \gamma = 1 \quad (0 < \sigma < \tfrac{3}{2});\quad \gamma = \sigma/(3-\sigma)(\tfrac{3}{2} < \sigma < 2);\quad \gamma = 2 \ (\sigma > 2) \tag{16}$$

$$d = 4\,:\quad \gamma = 1 \quad (0 < \sigma < \infty)\quad . \tag{17}$$

(Analogous formulae have been found for the critical behavior of a Bose-Einstein Condensation.[45])

We thus see that the strongest cooperative force corresponds to $\gamma = 1$ and a linear form for $h(x)$. As the cooperative force weakens (e. g., in one dimension) γ becomes larger and $h(x)$ becomes more curved.

We must now turn to an assessment of more realistic short-range forces which correspond to the usual experimental situation. The best established results are for the Ising model for which the new terms have confirmed the original conjecture of $\gamma = 5/4$. Of the remaining models it is convenient to separate those corresponding to classical vector interactions for which series can be derived using star lattice constants only. The Ising model of spin 1/2 $(I(\frac{1}{2}))$ can be regarded as a classical vector in one dimension, the planar classical Heisenberg (PCH) in two dimensions,[46] the classical Heisenberg (CH) in three dimensions;[47] following Stanley[43] we shall terminate this r dimensional classical vector interaction with the spherical model $(r \to \infty)$.

For PCH and CH numerical estimates of critical exponents are more tentative than for $I(\frac{1}{2})$. For example, earlier estimates[48] suggested $\gamma = 4/3$ for CH; as more estensive series became available[47] this was modified to 11/8 (which we have used); very recently it has been suggested[49] that this is still an underestimate and a value nearer to 1.39 is more likely to be correct. Nevertheless, even if the individual values shown in Table 4 are not completely correct, the general trend with increasing spin dimension r should be adequately represented.

Classical behavior is approached most closely for $r = 1$ $(I(\frac{1}{2}))$ for which $h(x)$ is nearly linear[50] and $\gamma = 1\frac{1}{4}$. As r increases, we move steadily away from classical

TABLE 4

SHORT RANGE FORCE IN THREE DIMENSIONS

	Critical Parameters			
	$I(\frac{1}{2})$	PCH	CH	Spherical
$\gamma (=4\beta)$	$1\ ^1/4$	$1\ ^5/16$	$1\ ^3/8$	2
α	$^1/8$	$^1/32$	$-\ ^1/16$	-1 (sp. heat $\sim \|t\|^{-\alpha}$)

behavior until as $r \to \infty$, we have $h(x) = (1+ax)^2$ and $\gamma = 2$ for the spherical model. It is interesting that for all the models listed in Table 4 the numerical values of the exponents correspond to $\delta = 5$. It is possible that this is a general property of short-range forces.[51]

For general spin less detailed information is available. For the Ising model it was suggested some years ago on the basis of terms then available that γ seemed to be independent of spin;[48] recent more extensive calculations have tended to confirm this conjecture.[49] For the Heisenberg model of spin 1/2 $\gamma \simeq 1.43$ (11/7) has been suggested;[30] the possibility is still open that γ is spin independent at least for finite s.

Equation (9) applies to the "critical region." Before we can use this with confidence, we must know how far the region extends, and for this we must calculate at least the first correction terms as we move away from the region. It is here that we refer to some ideas of Ninham,[52] who investigated the analytic behavior of the exact two-dimensional Onsager solution near the singularity. Ninham referred to 19th century investigations by Darboux of the form of the Taylor coefficients of a function $F(w)$ with the following analytic behavior near a singularity w_c:

$$F(w) = (w - w_c)^{-\gamma} f(w) + g(w) \tag{18}$$

where $f(w)$ and $g(w)$ are analytic at w_c. Equation (18) is satisfied by:

a) Known $I(\frac{1}{2})$ solutions in two dimensions.

b) The Spherical model.

c) Hypergeometric functions.

We may note that the singularity cannot be divided out. It is reasonable to assume as a first hypothesis that it applies also to other models, and for models which have been investigated there is good numerical support for the conjecture. For example, specific heats in the critical region for three-dimensional models on the FCC lattice are well fitted by the formulae[53,54]

$$C_H/R = \begin{cases} 1.091\,[(1-t')^{-1/8} - 1 - 1/8t' - 0.013t'^2] & (I(\tfrac{1}{2})) \\ 10.085\,[-(1-t')^{1/16} + 1 - 0.062t' - 0.011t'^2] & (CH) \end{cases} \qquad (19)$$

$$(t' = T_c/T) \quad (T > T_c)$$

so that we obtain an approximation to $f(w)$ and $g(w)$ in Eq. (19).

For the susceptibility extensive series like (5) and (6) enable a more detailed description to be given of the analytic behavior near the ferromagnetic and antiferromagnetic singularities.[55] (For the susceptibility of the two dimensional model exact calculations of the correlations[56] enable one to deduce that[57] Eq. (18) is satisfied in regard to the dominant singularity but not in regard to auxiliary singularities.)

Using Eq. (18) we can calculate the form of correction terms to (9), and if sufficiently long series expansions are available, it should be possible to estimate their numerical value. The correction terms are lattice dependent, and for the spherical model, several of them can be calculated exactly.[44] This serves as a useful guide to their order of magnitude.

I should like to close this section with a few brief remarks about correlations especially in nonzero field. The classical treatment is due to Ornstein and Zernike[58] who introduced the concept of a "direct" and "indirect" correlation function. Assuming that the direct correlation function has a sharp cut-off, their treatment leads to the formula for the correlation $\Gamma(\vec{\underline{R}})$ between two spins distant $\vec{\underline{R}}$ apart in zero field in d dimensions

$$\Gamma(\vec{\underline{R}}) = \exp - \kappa R / R^{d-2} \quad . \tag{20}$$

Here $\kappa = \kappa(t)$ behaves as $t^{1/2}$ when $T \to T_c$. Stell[59] has generalized this treatment to apply to long range forces; in fact, for any model in which the direct correlation function falls off in the same way as the force of interaction, he has shown that a generalized formula

$$\Gamma(\vec{\underline{R}}) = G(\kappa R)/R^{d-2} \quad (\kappa \to t^{1/2} \text{as } T \to T_c) \tag{21}$$

is valid, where $G(x)$ is an analytic function. Such models

he refers to as Ornstien-Zernike systems, and the spherical model and Bose-Einstein condensation both belong to this class.

To deal with deviations from Ornstein-Zernike behavior to be expected in other models, Fisher[60] suggested a generalization of (21) in the critical region as follows:

$$\Gamma(\vec{\underline{R}}) \simeq G(\kappa R)/R^{d-2+\eta}$$
$$\kappa \simeq t^{\nu} \quad . \tag{22}$$

It is easy to show that the exponents η and ν are not independent but satisfy the relation

$$\gamma = (2 - \eta)\nu \quad . \tag{23}$$

For the two dimensional Ising model exact calculations have shown that (22) is correct with $\nu = 1$ and $\eta = \frac{1}{4}$.

Two important questions arise in connection with the correlations:

a) Is the exponent ν determined by β and δ of Eq. (9)?

b) What happens in nonzero field? Can the critical behavior of the correlations be calculated in terms of the function $h(x)$ of Eq. (9)?

It is tempting to answer the second question in the affirmative. If we make the plausible assumption that (22) remains valid in nonzero field but that κ is now a function of T and H, we can show that[61] κ satisfies an equation analogous to M in Eq. (9),

$$H^2 = \kappa^{2\Delta/\nu} g(t\kappa^{-1/\nu}) \qquad (\Delta = \gamma + \beta) \tag{24}$$

where $g(x)$ can be calculated from $h(x)$.

However, the first question remains open. The scaling hypothesis of Kadanoff[35] suggests the following relation

$$d\nu = \gamma + 2\beta \tag{25}$$

which is satisfied by the known two-dimensional solutions for $I(\frac{1}{2})$ and by the spherical model for all ranges of force[44] in sufficiently low dimensions $(d \leq 4)$. It is fairly clear that in sufficiently high dimensions $(d > 4)$ the relation fails.[62] For the three-dimensional Ising model Fisher and Burford[60] suggested that it fails, but only just, $(\gamma = 5/4,\ \beta = 5/16,$ $\nu = 0.651)$. More recent investigations[63] at the University of Illinois have modified ν to 0.638 but not enough to satisfy Eq. (25) $(\nu = 0.625)$. Is this a genuine failure or an example of very slow convergence? The same group have also indicated[64] that the general relation Eq. (22) seems to fail for $I(\frac{1}{2})$ in three dimensions; it is apparently valid asymptotically for large κR but not as $\kappa \to 0$ for fixed R. Is this failure associated with the breakdown of Eq. (25)? Understanding of this aspect of the problem is still inadequate to supply an answer.

V. CONCLUSION

The above discussion may justifiably be criticized as being unduly subjective, and there are certainly a number of glaring omissions. No reference has been given to the remarkable series of exact solutions of ferroelectric models by Lieb.[65] This is not because I underestimate their significance, but I could not find a means of fitting these solutions into the general framework of the discussion. Similarly, I have ignored the elegant formalism[66] of the diagrammatic approach to high temperature expansions which does not make use of embeddings. This is because for several years this approach contributed little to our detailed knowledge of critical behavior, and it was only recently that the research group at Illinois has converted the method into an effective practical tool[67] (numerical results of this approach have been quoted[63, 64]). This approach may well prove subsequently to be of major importance in providing a proper justification for the scaling hypothesis.[68]

It is also valid to comment that I have taken a committed viewpoint in regard to the equation of state (9). This is partly because the equation made a number of predictions regarding critical exponents which were borne out by subsequent investigations, but also since I do not think one needs to abandon a simple description unless there is compelling evidence against it. One of the predictions of Eq. (9) which has given trouble is the symmetry of critical exponents above and below T_c; but so far the evidence is inconclusive[69] and convergence of low temperature series is notoriously slow.

The subject of rigorous analytical results is worthy of a paper in its own right, but I will draw attention briefly to one point of relevance to the problems considered here. This is contained in two papers by Yang and Lee[70] in 1952 who showed that for the Ising model all the zeros of the partition function are located on the unit circle in the plane of $y = (\exp -\beta mH)$. For $y < 1$ the partition function is analytic in y.[71] Critical properties arise from the condensation of zeros, and attention is focussed on the distribution of zeros on the unit circle. The distribution corresponding to Eq. (9) has been derived,[72] but although this provides a useful alternative point of view, it has not helped in the rigorous sense. (Incidentally, in one of the above papers Lee and Yang[70] used the Ising model to correct important errors in the description of condensation as currently presented by the Mayer theory.) Baker and Gaunt[73] have shown recently how the Yang-Lee theory can be used to provide information about critical exponents. They derive the following formula for the partition function:

$$\beta F = \frac{1}{2} \ln[\frac{1}{2}(1-\gamma^2)] + \int_0^\infty (\frac{1+w}{1+\tau^2 w}) \, d\varphi(w) \qquad (\tau = \tanh \beta mH). \quad (27)$$

Here $d\varphi(w) \geq 0$, $\int_0^\infty d\varphi(w) = \frac{1}{2}$, and φ depends on T but not on H. Assuming a gap index on the high temperature side, they use Eq. (27) to draw conclusions about the low temperature side.

I should like finally to refer to an attempt I have made recently to relate the Ising and Heisenberg models to the properties of self-avoiding walks on lattices.[28] It may be argued that this just reduces one untractable problem to another. In fact, for self-avoiding walks on lattices, <u>no</u>

analytic solutions have been obtained so far, yet nevertheless, I maintain that intrinsically they represent a simpler problem which should yield more readily to analytic treatment.[74]

There is a close correspondence between coefficients in high temperature Ising expansions and geometrical properties of self avoiding walks[75] as illustrated in Table 5. The "connective constant" μ differs from $1/w_c$ by about $2\frac{1}{2}\%$ in three dimensions, and the exponents behave in a parallel manner for both problems. However, to relate the problems more precisely, it is necessary to find a coherent scheme which lists contributions from graphs of higher cyclomatic number. These are related to near-neighbor contacts on the walk and the way in which they are ordered along it. If critical behavior is lattice independent as conjectured in Section IV, then the hydrogen peroxide lattice of coordination number 3 should already show the true critical exponents. Our attention is thus automatically focussed on the graphs with nodes of order three which were considered in Section III.

In summing up, we may observe that from the point of view of practical calculations of the behavior near the Curie point, we are well on the way to satisfying the needs of the

TABLE 5

ISING MODELS AND SELF AVOIDING WALKS IN THREE DIMENSIONS

Ising $(w = \tanh \beta J)$		Self avoiding walk
Specific heat		No. of polygons
$U = \Sigma a_n w^n$	$a_n \sim w_c^{-n} n^{-15/8}$	$\mu^n n^{-7/4}$
Susceptibility		No. of chains
$\chi_o = \Sigma b_n w^n$	$b_n \sim w_c^{-n} n^{1/4}$	$\mu^n n^{1/6}$
2nd Derivative of susceptibility		No. of overlaps of 2 chains
$\chi_o = \Sigma c_n w^n$	$c_n \sim w_c^{-n} n^{3\,3/8}$	$\mu^n n^{3\,1/2}$
Mean sq. coherence range		Mean sq. length of a chain
$\Sigma R^2 \Gamma(R)/\Sigma \Gamma(R) = \Sigma d_n w^n$	$d_n \sim w_c^{-n} n^{1.28}$	$\mu^n n^{6/5}$

experimentalist for most models of interest. Unifying features have been discovered which suggest that the critical behavior of a large variety of theoretical models can be described by a simple type of equation of state. But the rigorous mathematical theory needed to make the above developments "respectable" is still lacking.

ACKNOWLEDGMENTS

I am grateful to Professor L. F. Bates, F. R. S. and Professor N. Kurti, F. R. S. for helpful discussions in connection with the historical background in the Introduction, and to my colleages at King's College for their general help and cooperation.

REFERENCES

1. G. E. Uhlenbeck, Proc. Wash. Conf., (M. S. Green and J. V. Sengers, eds), N. B. S. Misc. Publ., 273, 3, (1966).

2. P. Curie, Ann. Chim. Phys., 5, 289 (1895); a parallel paper is P. Curie, J. de Phys., 4, 263 (1895).

3. J. Hopkinson, Phil. Trans. Roy. Soc. A, 1889, p. 443.

4. Nature, 58, 419 (1898); for a biographical sketch see J. Greig, John Hopkinson, Science Museum, London.

5. Bauer, Wied. Ann., 11, 394 (1880); W. F. Barrett, Phil. Mag., 47, 51 (1874); M. Faraday, Experimental Researches in Electricity, 1839-55, pp. 2343-7.

6. W. Gilbert, De Magnete Magneticisque Corporibus et de Magno Magnete Tellure Physiologia Nova, (translated by P. F. Mottelay, New York, 1893 and for the Gilbert Club, London, 1900) (1600), p. 66.

7. P. Weiss, J. de Phys., 6, 661 (1907).

8. P. Weiss and H. Kamerlingh Onnes, J. de Phys., 9, 555 (1910).

9. W. L. Bragg and E. J. Williams, Proc. Roy. Soc. A, 145, 699 (1934).

10. E. C. Stoner, Magnetism and Matter, Methuen, London 1934.

11. F. Cernuschi and H. Eyring, J. Chem. Phys., 7, 547, 1939.

12. H. D. Ursell, Proc. Camb. Phil. Soc., 23, 685 (1927).

13. J. E. Mayer, J. Chem. Phys., 5, 67 (1937); J. E. Mayer and Ph. G. Ackerman, J. Chem. Phys., 5, 74 (1937); J. E. Mayer and S. F. Harrison, J. Chem. Phys., 6, 87 (1938); S. F. Harrison and J. E. Mayer, J. Chem. Phys., 6, 101 (1938); J. E. Mayer, J. Phys. Chem., 43, 71 (1939).

13a. J. Yvon, Actualites Scientifiques et Industrielles, Paris, Herman, and Cie, No. 542, 1937.

13b. R. J. Riddell, "Contributions to the Theory of Condensation," Dissertation, Univ. of Michigan, 1951; R. J. Riddell and G. E. Uhlenbeck, J. Chem. Phys., 21, 2056 (1953); G. W. Ford, R. Z. Norman, and G. E. Uhlenbeck, Proc. Natl. Acad. Sci. Wash., 42, 203 (1956); G. W. Ford and G. E. Uhlenbeck, Proc. Natl. Acad. Sci. Wash., 42, 122, 529 (1956); 43, 163 (1957).

14. L. Onsager, Phys. Rev., 65, 117 (1944).

15. E. Ising, Z. Phys., 31, 253 (1925); for a historical review see S. G. Brush, Rev. Mod. Phys., 39, 883 (1967).

16. R. Peierls, Proc. Camb. Phil. Soc., 32, 477 (1936).

17. H. A. Kramers and G. H. Wannier, Phys. Rev., 60, 252, 263 (1941).

18. B. Kaufman, Phys. Rev., 76, 1232 (1949); B. Kaufman and L. Onsager, Phys. Rev., 76, 1244 (1949).

19. C. N. Yang, Phys. Rev., 85, 808 (1952); L. Onsager in discussion of Florence Conference, Il Nuovo Cimento (Suppl.) 6, 261 (1949).

20. J. G. Kirkwood, J. Chem. Phys., 18, 380 (1950).

21. See G. E. Uhlenbeck and G. W. Ford, Studies in Statistical Mechanics, Vol. 1, N. Holland Publ. Co., Amsterdam, 1962, p. 123, and Ref. 13b.

22. G. Polya, Acta Math., 68, 145 (1937).

23. R. W. Robinson, preprint; to be published, Enumeration of Nonseparable Graphs.

24. G. S. Rushbrooke, Disc. Far. Soc., 43, 7 (1967).

25. J. A. Leu, D. D. Betts, and C. J. Elliott, Can. J. Phys., 47, 1671 (1969); J. A. Leu, Phys. Letters, 29A, 641 (1969).

26. M. F. Sykes, J. W. Essam, B. R. Heap, and B. J. Hiley, J. Math. Phys., 7, 1557 (1966).

27. B. R. Heap, J. Math. Phys., 7, 1582 (1966); see also J. Nagle, J. Math. Phys., 7, 1588 (1966).

28. C. Domb, J. Phys., C3, 256 (1970).

29. G. A. Baker, H. E. Gilbert, J. Eve, and G. S. Rushbrooke, A Data Compendium of Linear Graphs with Application to the Heisenberg Model, BNL 50053 (T-460) (1967).

30. G. A. Baker, H. E. Gilbert, J. Eve, and G. S. Rushbrooke, Phys. Rev., 164, 800 (1967).

30a. C. Domb and M. F. Sykes, J. Math. Phys., 2, 63, (1961).

31. C. Domb, Proc. Wash. Conf., (M. S. Green and J. V. Sengers, eds.), N. B. S. Misc. Publ., 273, 29, (1966).

32. M. E. Fisher, Boulder Lectures in Theoretical Physics, 7C, 1 (1965); Rept. Progr. Phys., 30, 615 (1967).

33. J. W. Essam and M. E. Fisher, J. Chem. Phys., 38, 802 (1963); M. E. Fisher, Physics, 3, 255 (1967).

34. B. Widom, J. Chem. Phys., 43, 3892 (1965).

35. L. P. Kadanoff, Physics, 2, 263 (1966).

36. A. Z. Patashinskii and V. L. Pokrovskii, Zh. Eksper. Teor. Fiz., 50, 439; Sov. Phys. JETP, 23, 292 (1966).

37. C. Domb and D. L. Hunter, Proc. Phys. Soc., 86, 1147 (1965).

38. R. B. Griffiths, Phys. Rev., 158, 176 (1967).

39. See e.g., P. G. Watson, J. Phys. C2, 1883, 2158 (1969).

40. M. Kac, Phys. Fluids, 2, 8 (1959); M. Kac, G. E. Uhlenbeck, and P. C. Hemmer, J. Math. Phys., 4, 216, 229 (1963); 5, 60 (1964); G. A. Baker, Phys. Rev., 122, 1477 (1961); 130, 1406 (1963).

41. T. H. Berlin and M. Kac, Phys. Rev., 86, 821 (1952).

42. G. S. Joyce, Phys. Rev., 146, 349 (1966).

43. H. E. Stanley, Phys. Rev., 176, 718 (1968); 179, 570 (1969); Phys. Rev. Letters, 20, 589 (1968); J. Appl. Phys., 40, 1272 (1969); J. Phys. Soc. Japan, (Suppl.), 26, 102 (1969).

44. C. Domb and G. S. Joyce, to be published.

45. J. D. Gunton and M. D. Buckingham, Phys. Rev., 166, 152 (1968); M. J. Cooper and M. S. Green, Phys. Rev., 176, 302 (1968).

46. R. G. Bowers and G. S. Joyce, Phys. Rev. Letters, 19, 630 (1967).

47. H. E. Stanley and T. A. Kaplan, Phys. Rev. Letters, 16, 981 (1966); G. S. Joyce and R. G. Bowers, Proc. Phys. Soc., 88, 1053 (1966); P. J. Wood and G. S. Rushbrooke, Phys. Rev. Letters, 17, 307 (1966).

48. C. Domb and M. F. Sykes, Phys. Rev., 128, 168 (1962); J. Gammel, W. Marshall and L. Morgan, Proc. Roy. Soc. A, 275, 257 (1963).

49. M. Wortis, private communication, to be published.

50. D. S. Gaunt and C. Domb, J. Phys. C, 3, 1442 (1970).

51. J. D. Gunton and M. J. Buckingham, Phys. Rev. Letters, 20, 143 (1968).

52. B. W. Ninham, J. Math. Phys., 4, 679 (1963).

53. M. F. Sykes, J. L. Martin, and D. L. Hunter, Proc. Phys. Soc., 91, 671 (1967).

54. C. Domb and R. G. Bowers, J. Phys. C., 2, 755 (1969).

55. C. Domb and M. F. Sykes, to be published.

56. H. Cheng and T. T. Wu, Phys. Rev., 164, 719 (1967); T. T. Wu, Phys. Rev., 149, 380 (1966).

57. C. Domb, J. Phys. C., 2, 2433 (1969).

58. L. S. Ornstein and F. Zernike, Proc. Acad. Sci. Amst., 17, 793 (1914); Phys. Z., 19, 134 (1918); 27, 761 (1926).

59. G. Stell, Phys. Rev., 184, 135 (1969).

60. M. E. Fisher, J. Math. Phys., 5, 944 (1964); M. E. Fisher and R. J. Burford, Phys. Rev., 156, 583 (1967).

61. C. Domb, J. Phys. C., 2, 2433 (1969); L. S. Garcia-Colin and M. Ley Koo, Phys. Letters, 29A, 688 (1969).

62. C. Domb, Phys. Rev. Letters, 20, 1425 (1968).

63. M. A. Moore, D. Jasnow, and M. Wortis, Phys. Rev. Letters, 22, 940 (1969).

64. M. Ferer, M. A. Moore, and M. Wortis, Phys. Rev. Letters, 22, 1382 (1969).

65. E. H. Lieb, Phys. Rev. Letters, 18, 692, 1046 (1967); 19, 108 (1967); Phys. Rev., 162, 162 (1967).

66. R. Brout, Phys. Rev., 118, 1009 (1960); 122, 469 (1961); G. Horwitz and H. B. Callen, Phys. Rev., 124, 1757 (1961); F. Englert, Phys. Rev., 129, 567 (1963); C. Bloch and J. S. Langer, J. Math. Phys., 6, 554 (1965).

67. D. Jasnow, M. A. Moore, and M. Wortis, Phys. Rev., 185, 805 (1969).

68. See e. g., A. A. Migdal, Sov. Phys. JETP, in press; A. M. Polyakov, Sov. Phys. JETP, 28, 533 (1969).

69. See e. g., D. S. Gaunt, Proc. Phys. Soc., 92, 150 (1967).

70. C. N. Yang and T. D. Lee, Phys. Rev., 87, 404, 410 (1952).

71. J. L. Lebowitz and O. Penrose, Commun. Math. Phys., 11, 99 (1968); G. Gallavotti, S. Miracle-Sole, and D. W. Robinson, Phys. Letters, 25A, 493 (1967).

72. R. Abe, Prog. Theoret. Phys., 38, 72 (1967).

73. G. A. Baker, Phys. Rev. Letters, 20, 990 (1968); G. A. Baker and D. S. Gaunt, Phys. Rev., in press.

74. S. F. Edwards, Proc. Phys. Soc., 85, 613 (1965); H. Reiss, J. Chem. Phys., 47, 186 (1963); J. des Cloizeaux, J. Phys. Soc. Japan, (Suppl.), 26, 42, (1969); Z. Alexandrowicz, J. Chem. Phys., 46, 3789 (1967).

75. C. Domb, Advan. Chem. Phys., 15, 229 (1969).

PHASE TRANSITIONS IN FERROMAGNETS

Freeman J. Dyson

Institute for Advanced Study
Princeton, New Jersey

I. ONE-DIMENSIONAL ISING MODELS

A. I will not spend this hour going through the tedious details of the proof that the one-dimensional Ising ferromagnet with Hamiltonian

$$H = -\sum_{r>s} J(r-s)\mu_r\mu_s, \quad \mu_r = \pm 1 \tag{1}$$

has a phase transition when

$$J(n) = Jn^{-\alpha}, \quad J>0, \quad 1<\alpha<2 . \tag{2}$$

The details are in my papers.[1, 2] I want to talk mostly about the problems that remain unsolved.

Problem 1. Does the system (1), (2) have a phase transition when $\alpha = 2$?

This problem is of interest for several reasons. We know that the system is trivially ordered (frozen solid) at all temperatures when $\alpha \leq 1$, and that it is disordered at all temperatures when $\alpha > 2$ (see Ruelle[3]). In the range $1 < \alpha < 2$ there is ordering (spontaneous magnetization) at low T and disorder at high T. So the case $\alpha = 2$ is a singular point, the only case for which the behavior is unknown.

There are other reasons why Problem 1 is interesting. One would like to have necessary and sufficient conditions on the interaction function $J(n)$ for a phase transition to occur. According to Ruelle[3] the conditions

$$\Sigma\ J(n) < \infty\ , \quad \Sigma\ n\,J(n) = \infty \tag{3}$$

are necessary, but according to Dyson[2] not quite sufficient. According to Dyson[1] the conditions

$$\Sigma\ J(n) < \infty\ , \quad \Sigma\ (n^{\epsilon - 3}/J(n)) < \infty\ , \ \epsilon > 0\ , \tag{4}$$

are sufficient but not necessary. The case $J(n) = n^{-2}$ stands in the no-man's land between (3) and (4). In trying to guess what the correct necessary and sufficient conditions should be, it would be helpful (and almost indispensible) to know on which side of the fence $J(n) = n^{-2}$ ought to stand.

B. Anderson and his colleagues[4,5] have found a very elegant connection between the case $J(n) = n^{-2}$ of the Ising ferromagnet and a solid-state system called the Kondo problem. I am not the man to give you a lecture on the Kondo problem. For that you should invite Anderson. The interesting thing to me is that Anderson claims to have answered yes to Problem 1. Actually, after talking with Anderson for 3 hours, I do not believe he has settled the question with mathematical rigor. But he has done a calculation which is as good as, or better than, most calculations done by physicists, and which is in a certain sense asymptotically exact in the low-temperature limit. His result makes it at least highly plausible that the n^{-2} model has a phase transition.

C. The final thing I shall mention to justify being concerned about the n^{-2} model is a remarkably simple argument of Thouless.[6] Thouless shows that, if Anderson is right and there is a transition, the transition is of a new and strange kind. The argument of Thouless is intuitive and therefore probably correct.

Thouless Argument. Take $J(n) = Jn^{-2}$ and suppose the system has a spontaneous magnetization $m > 0$ at temperature T. Consider a long but finite chain of N spins. From the magnetized state we can derive a new state by reversing the direction of all spins to the right of a given spin j. This break in the ordering increases the energy by roughly

$$\Delta E = \sum_{r>j} \sum_{s \le j} 2Jm^2 (r-s)^{-2} \sim 2Jm^2 \log N , \tag{5}$$

assuming that the break is not close to either end of the chain. But the break can be made in roughly N different places, and therefore the broken states have collectively an additional entropy

$$\Delta S = \log N .$$

So the increase in free energy required to break the ordering is

$$\Delta F = \Delta E - T\Delta S = (2Jm^2 - T) \log N . \tag{6}$$

This means that the ordered state is stable only if

$$2Jm^2 > T . \tag{7}$$

For any temperature T, either (7) holds or else $m = 0$.

The result (7) is strange because we know that m^2 has to be a decreasing function of T. If there is a transition at some temperature T_c, then

$$m^2 > (T_c/2J) \tag{8}$$

for all $T < T_c$, while $m^2 = 0$ for $T > T_c$. There has to be a discontinuous jump in the magnetization at T_c. Such a jump has never been seen before in a ferromagnet, but it is not mathematically absurd. This behavior of course makes it all the more interesting to solve Problem 1. If one believes Anderson's asymptotic estimates, the magnetization stays high (above 85%) right up to the transition and then jumps to zero. We have also to solve

Problem 2. Make the Thouless argument rigorous.

D. Another light on the situation is thrown by some older calculations of Joyce[7] on spherical models with long-range interactions. These are systems with the Hamiltonian (1), (2) but with each μ_r a real variable subject only to the condition

$$\Sigma \mu_r^2 = \text{Constant} . \tag{9}$$

Spherical models are rather unphysical but have the great advantage that they can be solved analytically, and one may hope that they reflect in some ways the behavior of the corresponding Ising models.

The Joyce Phenomenon. Joyce is able to calculate the zero-field susceptibility $\chi(T)$, for temperatures T above the transition temperature T_c, for the spherical version of the system defined by (1), (2). He covers the range $1 < \alpha < 2$ over which a transition exists. He finds the "classical behavior"

$$\chi \sim (T - T_c)^{-1} \tag{10}$$

only for $1 < \alpha \leq 3/2$, whereas

$$\chi \sim (T - T_c)^{-\beta}, \quad \beta = (\alpha - 1)/(2 - \alpha), \tag{11}$$

for $3/2 \leq \alpha < 2$. There is a definite break at $\alpha = 3/2$ in the nature of the phase transition. So we have

Problem 3. Does the Joyce phenomenon occur also for Ising ferromagnets?

Or more generally,

Problem 4. Describe the nature of the phase transition for the Ising model (1).

Incidentally, the Joyce calculation gives a definite answer to Problem 1 for the spherical model. There is no transition for $\alpha = 2$. However, β in (11) tends to infinity as $\alpha \to 2$, so that the rise in susceptibility at the transition is becoming infinitely steep as we approach $\alpha = 2$. This is a delicate borderline case, and it could well behave differently for spherical and Ising models. Some numerical calculations by Nagle and Bonner[8] on short Ising chains give indications that the answer to Problem 3 may be yes.

II. HIERARCHICAL MODELS

A. Here is a picture of a hierarchical model.

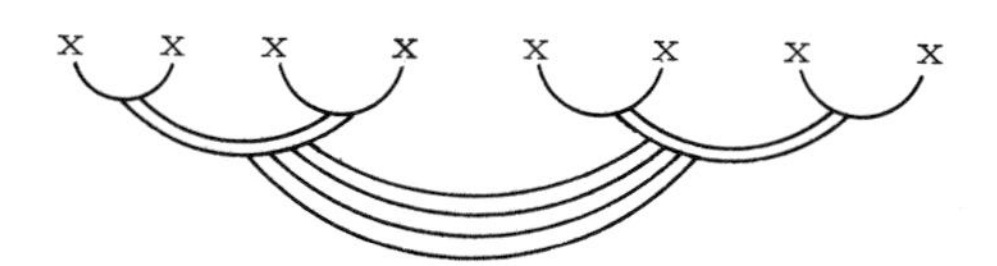

The crosses represent spins. The curves represent interactions. At the p'th level counting from the top, there is an interaction

$$-2^{-2p} b_p$$

between each pair of spins in two consecutive blocks of 2^{p-1} spins. The b_p are positive for a ferromagnetic system.

This model was invented as a smoothed-out version of the linear chain (1). The interactions are of roughly equal strength in both models if

$$2^{-2p} b_p = J(2^p) \; , \tag{12}$$

or equivalently

$$J(n) = n^{-2} b_{\log_2 n} \; . \tag{13}$$

Thus the case $J(n) = n^{-\alpha}$ corresponds to

$$b_p = 2^{p(2-\alpha)} \; , \tag{14}$$

and the critical case $J(n) = n^{-2}$ corresponds to

$$b_p = 1. \tag{15}$$

The interactions in the HM (Hierarchical model) are in each case weaker than the interactions in the corresponding LM (Linear model). So by using Griffiths inequalities, about which I shall speak in a moment, we can prove that a phase transition occurs in the LM if it occurs in the corresponding HM, but not conversely.

The following facts are known about the HM. There is a phase transition (Dyson[1]) if

$$\Sigma\, 2^{-p} b_p < \infty \quad , \quad \Sigma\, ((\log p)/b_p) < \infty \ . \tag{16}$$

There is no phase transition if the b_p are bounded. Thus there is a transition in the HM in the case (14) with $1 < \alpha < 2$, but there is no transition when $\alpha = 2$. The HM is in this way able to settle the question of a transition in the LM for $1 < \alpha < 2$ but is not able to settle the case $\alpha = 2$.

B. All that I have said so far about the HM refers to a system of Ising spins $\mu_r = \pm 1$. But the main motivation for studying the HM came from a different direction. One can prove the theorems on existence and nonexistence of phase transitions for the HM, and they remain true word-for-word, if the spins are quantum mechanical and the Ising interaction $\mu_r \mu_s$ is replaced by the Heisenberg interaction

$$(\vec{\sigma}_r \cdot \vec{\sigma}_s), \tag{17}$$

where each $\vec{\sigma}_r$ is a vector whose components are a set of 3 Pauli spin matrices. The Heisenberg HM has the property that its Hamiltonian is diagonalizable in terms of ordinary angular momentum operators. The statistical mechanics of the Heisenberg HM can therefore be handled as if it were a classical system, avoiding all the specifically quantum mechanical miseries which make the linear HM intractable. In particular, the Heisenberg HM with interaction (14) has a phase transition for $1 < \alpha < 2$. It is surprising that the change from Ising to Heisenberg spins makes so little difference in this case. Physical intuition, supported by some mathematical theorems of Mermin and Wagner,[9] leads one to expect that spontaneous magnetization would be much harder to achieve in a Heisenberg system than in an Ising system, because the Heisenberg system allows a gradual (Bloch wall) rotation of the spin direction while the Ising system does not.

I had hoped to use the Griffiths inequalities to deduce from the existence of phase transitions in the Heisenberg HM that a phase transition would also occur in the Heisenberg LM as it does in the Ising LM. This hope collapsed when Hurst and Sherman[10] discovered that the Griffiths inequalities are in general false for Heisenberg spins. I will mention later a conjecture which I believe may be true for Heisenberg spins and which would allow one to make the deduction from Heisenberg HM to Heisenberg LM. In the meantime, we still have not solved

Problem 5. Does the Heisenberg linear model analogous to (1), (2) have a phase transition for $1 < \alpha < 2$?

III. GRIFFITHS INEQUALITIES

A. The Griffiths inequalities were proved by Griffiths[11, 12] for any Ising system, independent of geometrical structure, with a Hamiltonian

$$H = -\sum_{1}^{N} J_{mn} \mu_m \mu_n \tag{18}$$

and all $J_{mn} \geq 0$. The inequalities are

Griffiths 1. $$\langle \mu_i \mu_j \rangle \geq 0 \; . \tag{19}$$

Griffiths 2. $$(\partial / \partial J_{mn}) \langle \mu_i \mu_j \rangle \geq 0 \; . \tag{20}$$

Griffiths 3. $$\langle \mu_i \mu_j \rangle \leq \sum_{k \neq i} \tanh(\beta J_{ik}) \langle \mu_j \mu_k \rangle \tag{21}$$

The brackets $\langle \; \rangle$ denote a mean value in thermal equilibrium at temperature $T = \beta^{-1}/k$. The most important and the richest in consequences is Griffiths 2, which says that every spin correlation is a monotonic function of every interaction strength.

If we add an external field B, the Hamiltonian becomes

$$H = -\Sigma J_{mn}\mu_m\mu_n - B\Sigma\mu_m . \tag{22}$$

Let

$$M = N^{-1} < \Sigma\mu_m > \tag{23}$$

be the mean magnetization. Then Griffiths 1 and 2 give

$$M \geq 0 , \tag{24}$$

$$(\partial M/\partial B) \geq 0 . \tag{25}$$

Quite recently Griffiths, Hurst, and Sherman[13] have proved by an immense tour de force of graphical analysis

<u>Griffiths 4.</u> $$(\partial^2 M/\partial B^2) \leq 0 . \tag{26}$$

This is the first step beyond the original work of Griffiths for ordinary Ising systems with binary interactions.

B. Meanwhile a great deal of powerful work has been done generalizing Griffiths 1, 2, and 3 to various more general types of systems. First, Kelly and Sherman[14] introduced the notion of a generalized Ising ferromagnet with the Hamiltonian

$$H = -\Sigma J_R \mu_R , \tag{27}$$

where R is any subset of $(1, 2, \ldots, N)$,

$$\mu_R = \prod_{m \in R} \mu_m , \tag{28}$$

and all $J_R \geq 0$. They proved the generalized Griffiths inequalities

<u>GG1.</u> $$\langle \mu_R \rangle \geq 0 . \tag{29}$$

<u>GG2.</u> $$(\partial / \partial J_S) \langle \mu_R \rangle \geq 0 . \tag{30}$$

Later, Ginibre[15,16] gave a much simpler and more illuminating proof of GG2; Kelly and Sherman[14] also proved

<u>GG3.</u> $$\langle \mu_R \rangle \leq \sum_S \tanh(\beta J_S) \langle \mu_R \mu_S \rangle \tag{31}$$

where r is any site in the set R, and S is summed over all sets containing r. However, the full generalization of Griffiths 4 would be

$$(\partial^2 / \partial J_R \partial J_S) \langle \mu_T \rangle \leq 0 , \tag{32}$$

and this was shown to be false by Kelly and Sherman.[14]

In the last few weeks there appeared an even more beautiful paper of Ginibre,[17] in which Griffiths 1 and 2 are proved for a variety of types of system by a new group theoretical method. A number of these latest results of

Ginibre were also proved by Sherman in a different way. However, it is still true in these latest investigations, as in the earlier ones, that the proofs are highly specific to the particular forms of interaction which are considered. Even among classical systems, the vast majority are not yet covered. So I add to the list of unsolved problems

Problem 6. To prove the Griffiths inequalities 1, 2, 3 in a way which applies uniformly to a wide class of classical ferromagnets.

C. Finally I mention briefly the extension of Griffiths inequalities to quantum systems. The most interesting case here is the Heisenberg system with Hamiltonian

$$H = -\sum_{1}^{N} J_{mn} (\vec{\sigma}_m \cdot \vec{\sigma}_n) \tag{33}$$

and all $J_{mn} \geq 0$. For this case Hurst and Sherman[10] proved

HG1. $$\langle \vec{\sigma}_i \cdot \vec{\sigma}_j \rangle \geq 0 , \tag{34}$$

but found the big surprise that

$$(\partial/\partial J_{mn}) \langle \vec{\sigma}_i \cdot \vec{\sigma}_j \rangle \geq 0 \tag{35}$$

is in general false. There is no known Heisenberg analogue to Griffiths 3. Later Gallavotti,[18] following the ideas of Ginibre,[16] generalized HG1 to a wider class of quantum systems. An outstanding task for the future is

Problem 7. To find inequalities going as far as possible in the direction of Griffiths 2 and 3 valid for Heisenberg ferromagnets. It seems intuitively obvious from a physical point of view that some monotonicity property along the lines of Griffiths 2 ought to hold for all ferromagnets, whether they are classical or quantum mechanical.

For my particular purpose of proving the existence of a phase transition in a linear Heisenberg model (Problem 5), knowing that the corresponding hierarchical model has a transition, it would be sufficient to have the following weakened form of HG2.

HG2 Conjecture. In the model (33) we have

$$(\partial/\partial J_{mn})\left\langle\left(\sum_{1}^{N} \sigma_j\right)^2\right\rangle \geq 0 . \tag{36}$$

This conjecture is at least true for a system of 3 spins, and so it does not fail on the Hurst-Sherman counter-example[10] which demolished (35). So my final unsolved problem is

Problem 8. Prove (36) for any Heisenberg system with all $J_{mn} \geq 0$.

These 8 problems should be enough to keep us busy for a while.

ACKNOWLEDGMENTS

I am grateful to Mark Kac and Colin Thompson for pulling me into this field and keeping me supplied with unsolved problems.

REFERENCES

1. F. J. Dyson, Commun. Math. Phys., 12, 91 (1969).

2. F. J. Dyson, Commun. Math. Phys., 12, 212 (1969).

3. D. Ruelle, Commun. Math. Phys., 9, 267 (1968).

4. P. W. Anderson and G. Yuval, Phys. Rev. Letters, 23, 89 (1969).

5. P. W. Anderson, G. Yuval, and D. R. Hamann, Exact Results in the Kondo Problem, II, Bell Lab. preprint, 1969.

6. D. J. Thouless, Phys. Rev., 187, 732 (1969).

7. G. S. Joyce, Phys. Rev., 146, 349 (1966).

8. J. F. Nagle and J. C. Bonner, Numerical Studies of the Ising Chain with Long-Range Ferromagnetic Interactions, Carnegie-Mellon Univ. preprint, 1969.

9. N. Mermin and H. Wagner, Phys. Rev. Letters, 17, 1133 (1966).

10. C. A. Hurst and S. Sherman, Phys. Rev. Letters, 22, 1357 (1969).

11. R. B. Griffiths, J. Math. Phys., 8, 478 and 484 (1967).

12. R. B. Griffiths, Commun. Math. Phys., 6, 121 (1967).

13. R. B. Griffiths, C. A. Hurst, and S. Sherman, Concavity of Magnetization of an Ising Ferromagnet, Univ. of Adelaide preprint, 1969.

14. D. G. Kelly and S. Sherman, J. Math. Phys., 9, 466 (1968).

15. J. Ginibre, Phys. Rev. Letters, 23, 828 (1969).

16. J. Ginibre, Correlations in Ising Ferromagnets, Orsay preprint, 1969.

17. J. Ginibre, General Formulation of Griffiths Inequalities, Orsay preprint, 1969.

18. G. Gallavotti, A Generalization of the First Griffiths Inequality to the Heisenberg Model, Rockefeller Univ. preprint, 1969.

FROM THE MEAN FIELD APPROXIMATION TO THE METHOD OF RANDOM FIELDS

A. J. F. Siegert

Northwestern University
Evanston, Illinois

The method of random fields is based on a representation of the partition function as an average over Gaussian random variables or random functions. This representation replaces all or part of the interaction between the spins of an Ising model, or between the molecules of a fluid, by impressed fields, with subsequent averaging over these fields with a Gaussian probability distribution specified by the interaction. The representation was first suggested by M. Kac.[1] It is exact for a class of interactions which is sufficiently large to be of physical interest. This survey will be restricted to the Ising model of ferromagnetism and to classical gas-liquid systems, omitting applications to systems with Coulomb interaction and the generalization to quantum statistical mechanics.

The mean field approximations replace all or part of the interaction by a spatially homogeneous impressed field. I will restrict myself here to the Weiss theory of ferromagnetism and the van der Waals theory with Maxwell (equal area) construction of the equation of state of a gas-liquid system, and sketch the developments which lead to the present understanding of these approximations as limit theorems. I will show how these mean field approximations are contained as zeroth approximations in the method of random fields.

The Weiss approximation replaces the system of interacting spins by a system of noninteracting spins, each in the same magnetic field F. The magnetization of this system is then easily obtained, let it be

$$M = M\left(\frac{\mu F}{kT}\right), \tag{1}$$

where μ is the magnetic moment associated with each spin, k the Boltzmann constant, and T the temperature. The mean field F is in turn determined by the applied field H and the magnetization M:

$$F = H + \alpha M . \tag{2}$$

These two equations for M and F provide a qualitative description of ferromagnetism and the existence of a Curie temperature and give the Curie-Weiss law for the susceptibility. The value of α needed to obtain the right order of magnitude for Curie temperatures could not be understood by magnetic interaction and was explained only later by Heisenberg's exchange interaction.

The van der Waals approximation replaces a system of particles interacting with a strong, short range repulsion $u(\vec{r}_i - \vec{r}_j)$, and a longer range, weaker attraction $-\upsilon(\vec{r}_i - \vec{r}_j)$ by a system of particles interacting only with interaction u (the "reference system") in a spatially homogeneous average potential.

This results in an equation for the pressure P(v) as function of the volume per particle v:

$$P(v) = P_h(v) - \frac{\upsilon_o}{2v^2} \tag{3}$$

where $P_h(v)$ is the pressure of the reference system and

$$\upsilon_o = \int \upsilon(\vec{r}) d^D r \tag{4}$$

where D is the dimensionality of the model and the integral is taken over all space. We will refer to Eq. (3) as the van der Waals equation, although in the equation usually referred to as van der Waals equation, $P_h(v)$ is replaced by an approximation which is exact only for the one-dimensional gas of hard rods.

This equation gives a semiquantitative understanding of the isotherms of the non-ideal gas, and of the critical temperature. At temperatures below the critical temperature the well-known van der Waals loops appear, so that Eq. (3), considered as an equation for the volume as function of the pressure, consists of three branches for a range of pressures. The equilibrium value of the volume is then chosen as the value with the lowest Gibbs free energy. The change from one branch to the other, at the pressure for which two branches give the same Gibbs free energy, is interpreted as condensation, and the Maxwell or "equal area" construction determines the condensation pressure.

The attitude of physicists toward the van der Waals equation has changed several times since its birth in 1873. After the Ursell-Mayer expansion in the mid-thirties, the orthodox view was that the van der Waals approximation was merely an extrapolation from the first two terms of the virial series, and the equal area construction an ex post facto introduction of thermodynamics, which would not be necessary if one could actually evaluate the partition function exactly, and obtain from it the pressure in the thermodynamic limit (number of particles $N \to \infty$, volume $V \to \infty$, with $V/N = v$ fixed).

There was, however, at that time no proof for any classical system that the van der Waals loops would not result even from an exact calculation of the Gibbs integral and that phase transitions were contained in the fundamental formalism of equilibrium statistical mechanics; and some very respectable physicists expressed a minority view to the contrary still in the late thirties.

In the forties, however, the famous papers of Onsager[2] and of van Hove[3] appeared. The former showed rigorously the phase transition (Curie point) of the two-dimensional Ising model with nearest neighbor interaction without magnetic field. The latter proved the impossibility of van der Waals loops in the thermodynamic limit for systems of particles with hard core repulsive interaction and finite range attraction. Van Kampen found a flaw in van Hove's proof ten years later, but the theorem is correct, and more general theorems of this type have since been proved by Ruelle[4] and by Fisher.[4]

The mean field theories were also contradicted by the proofs of Rushbrook and Ursell[5] (for one-dimensional ordered assemblies) and of van Hove[6] (for the one-dimensional gas) that these systems do not have phase transitions if the range of interaction is finite.

All this seemed to close the case, but I remember Professor Uhlenbeck saying at that time, that is in the late forties, that he was unwilling to believe that a theory which gave a qualitative and semiquantitative description of the isotherm, including the phase transition, did not have some nucleus of truth in it.

The idea that this nucleus of truth would be a limit theorem stating that the mean field approximation should be <u>exact</u> in the limit of infinitely weak interaction of infinite range, must have been in people's minds; but I think it was first put in print by Brout[7] in his 1960 Ising model paper, in which he also started the investigation of the neighborhood of this limit.

The first rigorous proof of such a limit theorem was obtained by Baker[8] and Kac[9] in 1961, independently and by different methods. They solved exactly the one- dimensional Ising model, without magnetic field, with interaction energy

$$\mp c\gamma e^{-\gamma|k-\ell|} \quad \text{for} \quad \left\{ \begin{matrix} \text{parallel} \\ \text{antiparallel} \end{matrix} \right\} \quad \text{spins} \tag{5}$$

located at lattice sides k and ℓ, respectively. For $c > 0$, the Weiss theory of ferromagnetism is found to be correct in the limit $\gamma \to 0$. It is of course understood that the thermodynamic limit is taken first, before letting γ approach zero.

Uhlenbeck, Kac, and Hemmer[10] in three papers in 1963-64 investigated a one-dimensional model of particles which interact with hard core repulsion and an exponential attraction

$$-v(x_i - x_j) = c\gamma e^{-\gamma|x_i - x_j|} \tag{6}$$

where c is a positive constant. They proved that the van der Waals isotherm <u>with</u> Maxwell construction is the exact limit $\gamma \to 0$. Again, it is understood that the thermodynamic limit is taken first.

In both cases, one-dimensional Ising and one dimensional gas with exponential potential, there is no phase transition for $\gamma > 0$. This is a slight extension of the theorems of Rushbrook and Ursell,[5] and van Hove[6] concerning the nonexistence of phase transitions in one-dimensional systems with finite range interaction.

Kac[9] and Uhlenbeck, Kac and Hemmer[10] used for their exact solutions the method of random fields, which was suggested first by Kac[1] in 1957. One special simplifying feature used in their calculations, the Markov property of the random field, is not easily generalized to two or more dimensions, and interaction other than exponential. However, van Kampen[11] in 1964 showed heuristically and Lebowitz and Penrose[12] in 1966 rigorously <u>for any number of dimensions</u> that the van der Waals theory <u>with Maxwell construction</u> is the exact limit for particles which interact with strong short range repulsion, and infinitely weak, infinitely long range, integrable attraction. Specifically, the attractive part of the interaction is assumed to be

$$- v = \gamma \phi(\gamma^{1/D} r) \geq 0 \qquad (7)$$

where D is the number of dimensions, r the distance between particles, and $\int \phi(r)\, d^D r$ is finite.

Since the Ising model is equivalent to a lattice gas, this implies the validity of the Weiss theory in the limit of infinitely weak, infinite range, integrable interaction.

The method of random fields is fundamentally based on the following integral identity:

For n arbitrary complex numbers μ_k and for a positive definite n by n matrix υ

$$e^{\frac{1}{2}\sum_{k,\ell}\mu_k\mu_\ell\upsilon_{k\ell}} = \tag{8}$$
$$= (2\pi)^{-\frac{n}{2}}(\det\upsilon)^{-\frac{1}{2}}\int_{-\infty}^{\infty}\cdots\int e^{-\frac{1}{2}\sum_{k,\ell}\varphi_k\varphi_\ell(\upsilon^{-1})_{k\ell}+\sum_k\mu_k\varphi_k}\prod_j d\varphi_j .$$

This integral identity is easily proved by diagonalization of the matrix υ. It is convenient to think of the function

$$w_n \equiv (2\pi)^{-n/2}(\det\upsilon)^{-1/2}\exp\{-\frac{1}{2}\sum_{k,\ell}\varphi_k\varphi_\ell(\upsilon^{-1})_{k\ell}\} \tag{9}$$

as a probability density for Gaussian random variables, since $w_n \geq 0$, and, as one sees by choosing all $\mu_j = 0$ in Eq. (8), w_n is normalized to unity. The right-hand side of Eq. (8) can then be written as an average with respect to this probability density

$$\exp\{\frac{1}{2}\sum_{k,\ell}\mu_k\mu_\ell\upsilon_{k\ell}\} = \langle\exp\sum_k\mu_k\varphi_k\rangle_{av} . \tag{10}$$

Expanding both sides of Eq. (8) in the variables μ_k, one obtains average and covariance

$$< \varphi_j >_{av} = 0 \tag{11}$$

$$< \varphi_k \varphi_\ell >_{ak} = v_{k\ell}. \tag{12}$$

We note for later that Eq. (10) can therefore be written in the form

$$\exp \{ \frac{1}{2} \sum_{k,\ell} \mu_k \mu_\ell < \varphi_k \varphi_\ell >_{av} \} = < \exp \sum_k \mu_k \varphi_k >_{av} \tag{13}$$

Now consider an Ising model of n sites without magnetic field. Let $k, \ell \ldots$ designate position vectors of the sites on a D-dimensional lattice. Let $\mu_k, \mu_\ell \ldots = \pm 1$ be the "spin" variables associated with sites k, ℓ and let the interaction energy of a pair of spins at sites k and ℓ $(k \neq \ell)$ be $-kTv_{k\ell}\mu_k\mu_\ell$. We assume homogeneity, i. e., we let $v_{k\ell}$ depend only on the vectorial distance $k - \ell$, and for simplicity we consider only the purely ferromagnetic case $v_{k\ell} > 0$. It is also usually convenient to approach the thermodynamic limit $(n \to \infty)$ through a sequence of interactions with Born-von Kármán periodicity conditions.

The partition function is then

$$Q_n = \sum_{(\mu)} \exp\left[\frac{1}{2} \sum_{\substack{k,\ell \\ k \neq \ell}} \mu_k \mu_\ell v_{k\ell}\right] \tag{14}$$

where $\sum_{(\mu)}$ means sum over all $\mu_k = \pm 1$. The terms of this sum are of the form of the left-hand side of Eq. (10), except for the terms with $k = \ell$ in the sum in the exponent. These do not occur in the physical problem, and we are free to complete the definition of the matrix $v_{k\ell}$ by letting all the diagonal elements be equal to v_{oo}. If v_{oo} is chosen such as to make the matrix $v_{k\ell}$ positive definite, one obtains from Eq. (10) the representation

$$Q_n = \sum_{(\mu)} \langle \exp \sum_k \mu_k \varphi_k \rangle_{av} \, e^{-n v_{oo}/2} . \tag{15}$$

The order of summation and averaging can be interchanged, and one obtains the partition function of an Ising paramagnet, with a magnetic field proportional to φ_k at each site, averaged over these fields with the probability density w_n, given by Eq. (9). The sum is now easily carried out, and one obtains

$$Q_n = 2^n e^{-n v_{oo}/2} \langle \prod_k \cosh \varphi_k \rangle_{av} . \tag{16}$$

A range parameter is introduced explicitly by choosing for $\upsilon_{k\ell}$ the form

$$\upsilon_{k\ell} = \nu\gamma\rho_{k\ell} \tag{17}$$

with

$$\rho_{k\ell} = \rho(\gamma^{1/D}(k-\ell)) \tag{18}$$

for infinite models. For finite models, one usually imposes Born-von Kármán conditions and chooses a periodic interaction for which $n \to \infty$ approaches $\rho(\gamma^{1/D}(k-\ell))$ for any finite distance $|k-\ell|$. (There is a slight difference in notation between papers on Ising models and papers on fluids: In the latter the reciprocal range is γ while here it is $\gamma^{1/D}$.) One introduces a coupling constant J and defines ν by

$$\nu = \frac{J}{kT} . \tag{19}$$

The diagonal elements of ρ are chosen to be $\rho_{kk} = \rho(0) = 1$. With this notation one then has, writing the average explicitly,

$$Q_n = 2^n e^{-\frac{n\nu\gamma}{2}} (2\pi)^{-n/2} (\det\rho)^{-1/2} \cdot \tag{20}$$

$$\cdot \int_{-\infty}^{\infty} \cdots \int e^{-\frac{1}{2} \sum_{k,\ell} x_k (\rho^{-1})_{k\ell} x_\ell} \prod_k \cosh\sqrt{\nu\gamma}\, x_k dx_k$$

provided that the matrix ρ is positive definite.

For the one-dimensional model with exponential interaction (5) Kac[9] used the fact that the x_k's in that case form a Markov chain to evaluate this integral. This Markov property means that the probability density

$$W_n = (2\pi)^{-n/2}(\det \rho)^{-1/2} \exp\{-\frac{1}{2}\sum_{k,\ell} x_k(\rho^{-1})_{k\ell}\, x_\ell\} \tag{21}$$

factors into

$$W_n = W_1(x_1)\, P(x_1, x_2)\, P(x_2, x_3) \ldots P(x_{n-1}, x_n) \tag{22}$$

with

$$P(x_1, x_2) = W_2(x_1, x_2)/W_1(x_1) \tag{23}$$

(Born-von Kármán conditions are not imposed in this case, since they would interfere with the Markov property.) The integration over $x_2, x_3, \ldots, x_{n-1}$ then results in the (n-2)-fold iterate of the kernel

$$P(x_1, x_2)\, [\cosh(\sqrt{\nu\gamma}\, x_1) \cosh(\sqrt{\nu\gamma}\, x_2)]^{1/2}.$$

The problem of evaluating the partition function is thereby reduced to the solution of an integral equation with this kernel.

Kac,[13-15] Helfand,[13] and Thompson[14] have investigated some two-dimensional models for which the Boltzmann factor can be represented as the average over a functional of a multidimensional Markov process. This procedure leads already for two-dimensional models to an integral equation with the number of variables equal to the number of rows of lattice sites. By perturbation methods Kac and Thompson[14] obtained corrections to the Weiss-Curie temperature and bounds for the free energy through the critical region defined below. These bounds are not yet sufficiently strong to prove the critical region resummation.

I would like to describe here how the Weiss approximation appears as zeroth approximation of (20) for the wider class of models defined by the restrictions on the matrix ρ, and to sketch attempts to evaluate the free energy near the Weiss limit. The natural first step is to look for maxima of the integrand in Eq. (20). At temperatures above the Curie temperature of the Weiss approximation, T_w, the integrand has only one maximum: all $x_k = 0$. Below T_w there are more maxima. The two largest maxima are equal and are located at the points: all $x_k = x$ and all $x_k = -x$, where x is just the value obtained from the equation for the Weiss field. Replacing the integral in Eq. (20) by the largest value of the integrand yields the Weiss approximation.

Taking $\prod_j \cosh(\sqrt{\nu\gamma}\, x_j)$ into the exponent as $\sum_j \ln\cosh\sqrt{\nu j}\, x_j$ in the integrand of Eq. (20) and expanding around the maximum, or around the two largest maxima to second order in the variables x_k, one obtains Brout's original correction term to the Weiss approximation for the free energy, a term of order γ.

One can obtain an upper bound on the partition function[16] (i. e., a lower bound on the free energy) by adding and subtracting in the exponent the term $\nu\gamma \sum_k x_k^2/2\alpha$. If $\ln\cosh\sqrt{\nu\gamma}\, x_k^2 - \nu\gamma x_k^2/2\alpha$ is replaced by its upper bound, the integral can be carried out explicity and is finite provided that α is chosen such that $\alpha/\nu\gamma$ is larger than the largest eigenvalue of ρ which requires $\alpha > T_w/T$. The remaining freedom in the choice of α is used to minimize the resulting upper bound on Q_n. Together with the known lower bound[17] on Q_n, this serves to show that the error of Brout's approximation to the free energy is $0(\gamma^2)$ for any fixed temperature above T_w. In the critical region, defined below, the difference between the bounds approaches zero for $\gamma \to 0$, but is of order of the square of the width of the critical region and thus not yet small enough to prove the critical region resummation. The complete expansion in ascending orders in γ was carried out, for any fixed temperature $T > T_w$, by Siegert and Vezzetti.[18] This expansion is actually obtained from a modification of the representation (20). We note first that, trivially, Q_n can also be written in the form

$$Q_n = e^{\frac{n}{2}(1-\nu\gamma)} \sum_{(\mu)} e^{-\frac{1}{2}\sum_{k,\ell}(\delta_{k\ell} - \nu\gamma\rho_{k\ell})\mu_k\mu_\ell} . \tag{24}$$

As long as the matrix $(\delta_{k\ell} - \nu\gamma\rho_{k\ell})$ is positive definite, that is for temperatures $T > T_w$, one can use random variables ξ_k which have this matrix as covariance to represent Q_n as an average over these variables:

$$\begin{aligned} Q_n &= e^{\frac{n}{2}(1-\nu\gamma)} \sum_{(\mu)} \langle e^{i\sum_k \xi_k \mu_k} \rangle_{av\xi} \\ &= 2^n e^{\frac{n}{2}(1-\nu\gamma)} \langle \prod_k \cos \xi_k \rangle_{av\xi} . \end{aligned} \tag{25}$$

The expansion obtained in this way has the feature that at $T = T_w$ almost all terms become infinite. Already when $T - T_w$ is small in some order of γ, the expansion becomes poor because most of the terms have denominators which vanish with some power of $T - T_w$. There is a "critical region" of temperatures, in which the first few terms become of the same order in γ. For typical 1-, 2-, and 3-dimensional models this critical region is of order $\gamma^{2/3}$, $\gamma \ln \frac{1}{\gamma}$, and γ, respectively. By "typical" we mean the model is not of the type of the long range models considered by Dyson,[19] but that $\sum_{\ell} \rho_{k\ell} \ell^2 / \sum_{\ell} \rho_{k\ell}$ is of order $\gamma^{-2/D}$ for small γ.

The failure of the expansion in the critical region is the fault of the straightforward γ expansion. Intuitively the thermodynamic functions should be analytic in T at T_w for finite γ.

Brout noticed the failure of the straight γ expansion and abandoned it in favor of a method which he called sphericalization, to obtain a result which avoids the failure at T_w. Horwitz and Callen[20] obtained essentially the same result by resummation of an expansion in reciprocal temperature. This result was also obtained by Muehlschlegel and Zittartz[17] who applied a variational technique to the integral (20). Siegert and Vezzetti[18] used a simple modification of (24) starting from

$$Q_n = e^{\frac{n}{2}(\alpha - \nu\gamma)} \sum_{(\mu)} e^{-\frac{1}{2} \sum_{k,\ell} (\alpha \delta_{k\ell} - \nu\gamma\rho_{k\ell}) \mu_k \mu_\ell} . \qquad (26)$$

One obtains a representation of Q_n as an average analogous to Eq. (25), in principle without restriction, because one is free to choose the parameter α so as to make the matrix $\alpha\delta_{k\ell} - \nu\gamma\rho_{k\ell}$ positive definite. In this way the "sphericalized" result of Brout is obtained as a beginning of a systematic expansion.

Thompson, Siegert, and Vezzetti[21] then showed that this result is obtained for the critical region by resumming the most divergent terms of the straight γ expansion, provided that there is a phase transition. (From Griffith's[22] work one knows that there is a phase transition for the typical model in two or more dimensions). The terms left over after resummation are individually of higher order in γ than the resummed terms.

* * *

The application of the method of random fields to the fluid requires the generalization of the definition of random fields from random variables to random functions. For the Ising model the random fields needed to be defined only at the lattice sites. Now we need random <u>functions</u> defined on a continuum.

These can be defined as follows:

Given any n points $\vec{r}_1, \vec{r}_2, \ldots, \vec{r}_n$, for arbitrary n, the function values $\varphi(\vec{r}_1), \varphi(\vec{r}_2), \ldots, \varphi(\vec{r}_n)$ are Gaussian random variables.

Then we have, by taking in Eq. (13) all $\mu_k = 1$,

$$\langle e^{\sum_j \varphi(\vec{r}_j)} \rangle_{av} = e^{\frac{1}{2} \sum_{jk} \langle \varphi(\vec{r}_j)\varphi(\vec{r}_k) \rangle_{av}} \tag{27}$$

Let the attractive part of the potential for a pair of particles at $\vec{r}_j, \vec{r}_k$ be $-\upsilon(\vec{r}_j - \vec{r}_k)$. If $\beta\upsilon(\vec{r}_j - \vec{r}_k)$ is the covariance of a Gaussian random function, that is

$$\beta\upsilon(\vec{r}_j - \vec{r}_k) = \langle \varphi(\vec{r}_j)\varphi(\vec{r}_k) \rangle_{av}, \tag{28}$$

then the Boltzmann factor for the attractive part can be written as the l. h. s. of (27) with a correction factor to eliminate the terms with $j = k$:

$$e^{\frac{\beta}{2} \sum_{j,k=1}^{N}{}' \upsilon(\vec{r}_j - \vec{r}_k)} = \langle e^{\sum_{j=1}^{N} \varphi(\vec{r}_j)} \rangle_{av} \, e^{-\frac{\beta}{2} N\upsilon(0)} . \tag{29}$$

The condition for Eq. (28) is positive definiteness of υ considered as the kernel of an integral equation (if one wants to work with real-valued random functions); and $\upsilon(0)$ has to be finite.

In the papers on fluids, the representation by random fields is not employed for the strong short range repulsion u, because this does not seem to be useful. The rule of the game is that everything about the reference system, i. e., the system with interaction u only, is assumed to be known (molecular distribution functions as well as thermodynamic functions). The papers of Zittartz,[23] and Jalickee, Siegert, and Vezzetti,[24] who use the method of random fields, as well as Brout and Coopersmith,[25] Hemmer,[26] and Lebowitz, Stell, and Baer,[27] who use diagram techniques, are aimed at obtaining the perturbation of the reference system by the weak long range attraction.

Using Eq. (29) to represent the Boltzmann factor for the attractive part of the interaction and introducing the abbreviation $S_N(\vec{r}_1, \vec{r}_2, \ldots, \vec{r}_N)$ for the Boltzmann factor of the reference system, one writes the Gibbs integral for N particles in a volume V in the form

$$Z_N = \frac{1}{N!} e^{-\frac{N\beta v(o)}{2}} \left\langle \int_V \ldots \int S_N(\vec{r}_1, \vec{r}_2, \ldots, \vec{r}_N) e^{\sum_{j=1}^{N} \varphi(\vec{r}_j)} \prod_j d^D r_j \right\rangle_{av}. \quad (30)$$

Except for the factor $\exp[-N\beta v(o)/2]$, this is the Gibbs integral for the reference system in a field $-\beta^{-1}\varphi(\vec{r})$, averaged over all such fields with a Gaussian probability distribution in the space of the fields.

The grand canonical partition function

$$\varnothing(z) \equiv \sum_{N=0}^{\infty} z^N Z_N \tag{31}$$

is then

$$\varnothing(z) = \langle \varnothing_h(\tilde{z}\,;\varphi(\vec{r})\rangle_{av} \tag{32}$$

where

$$\tilde{z} \equiv z \exp(-\beta v(o)/2) \tag{33}$$

and

$$\varnothing_h(\tilde{z};\varphi(\vec{r})) \equiv \sum_{N=0}^{\infty} \frac{\tilde{z}^N}{N!} \int_V \cdots \int S_N(\vec{r}_1, \vec{r}_2, \ldots, \vec{r}_N) e^{\sum_{j=1}^{N} \varphi(\vec{r}_j)} \prod_{j=1}^{N} d^D r_j \tag{34}$$

is the grand canonical partition function for the reference system in a field $-\beta^{-1}\varphi(\vec{r})$, i.e., the grand canonical partition functional.

The exact solution of the problem for a one-dimensional system of particles with hard core repulsion and exponential attraction [given by Eq. (6)] by Uhlenbeck, Kac and Hemmer[10] was possible because of the Markov property of Gaussian random functions of one variable with exponential covariance, and by the ordering possible for particles with hard core repulsion in the one-dimensional case.

To exhibit the van der Waals limit as the zeroth approximation, it is convenient to use the representation of the random functions

$$\varphi(\vec{r}) = \sum_{\sigma} c_{\sigma} \sqrt{\beta \upsilon_{\sigma}}\, u_{\sigma}(\vec{r}) \tag{35}$$

where υ_{σ} and $u_{\sigma}(\vec{r})$ are the eigenvalues and eigenfunctions, respectively, of the homogeneous integral equation with kernel $\upsilon(\vec{r} - \vec{r}\,')$ in the domain V, and the variables c_{σ} are independent Gaussian random variables with probability density $(2\pi)^{-1/2}\exp(-c_{\sigma}^{2}/2)$. With Born-von Kármán periodicity conditions the eigenvalues and eigenfunctions are simply the Fourier coefficients of $\upsilon(\vec{r})$, and trigonometric functions, respectively. With $\upsilon(\vec{r})$ assumed non-negative, the largest eigenvalue is

$$\upsilon_{o} = \int_{V} \upsilon(\vec{r})\, d^{D}r \, , \tag{36}$$

and the corresponding normalized eigenfunction is constant

$$u_{o} = V^{-1/2} \, . \tag{37}$$

Using this representation, one can write the average over the fields more explicitely as

$$\emptyset(\tilde{z}) = \int_{-\infty}^{\infty} \cdots \int \prod_{\sigma} e^{-\frac{1}{2} c_{\sigma}^{2}} \frac{dc_{\sigma}}{\sqrt{2\pi}} \emptyset_{h}(\tilde{z}\,; \varphi(\vec{r})) \tag{38}$$

where $\emptyset_h(\tilde{z}\,;\varphi(\vec{r}))$ is now a function of the variables c_{σ} through Eq. (35).

The equations for the extrema of the integrand are very complicated, but they have some trivial solutions c_{σ}^{*}, where $c_{\sigma}^{*} = 0$ for $\sigma \neq 0$, and c_{o}^{*} is a root of the equation

$$c_0 = \partial \ln \emptyset_h(\tilde{z} \exp(\sqrt{\frac{\beta v_o}{V}}\, c_o))/\partial c_o\,. \tag{39}$$

If one defines y by

$$y = \tilde{z} \exp(\sqrt{\frac{\beta v_o}{V}}\, c_o), \tag{40}$$

this equation becomes

$$\ln y - \ln \tilde{z} = \beta v_o \rho_h(y) \tag{41}$$

where $\rho_h(y)$ is the density of the reference system at fugacity y. Approximating the integral in Eq. (38) by the largest value of the integrand, one obtains for the isotherm the van der Waals result with Maxwell construction, since in this approximation the pressure becomes

$$\frac{1}{V} \ln \phi(z) \cong \frac{1}{V} \ln \phi^{(0)}(z) \equiv \frac{1}{V} \{ \ln \phi_h(y^*) - \frac{1}{2} c_o^{*2} \}$$

$$= \frac{1}{V} \ln \phi_h(y^*) - \frac{1}{2\beta v_o} (\ln y^* - \ln \tilde{z})^2 = \frac{1}{V} \ln \phi_h(y^*) - \frac{\beta v_o}{2} \rho_h^2 (y^*) \quad (42)$$

where y^* is that root of Eq. (41) for which the right-hand side assumes its largest value. As one would expect, the Maxwell construction in the grand canonical ensemble, that is the graphical solution of Eq. (41), is very similar to the Weiss construction, with $\ln \tilde{z}$ corresponding to the applied field, $\ln y$ to the local field, and ρ_h to the magnetization of the paramagnet.

This result suggests an expansion in the spatially inhomogeneous part

$$\psi(\vec{r}) \equiv \sum_{\sigma \neq 0} c_\sigma \sqrt{\frac{\beta v_\sigma}{V}}\, u_\sigma(\vec{r}) \quad (43)$$

of the random fields, with the average performed in two steps, first over ψ and then over the spatially homogeneous part of φ, which can be combined with the fugacity so that

$$\phi(z) = \langle \phi_h(\tilde{z}; \varphi(\vec{r})) \rangle_{av} = \langle\langle \phi_h(\tilde{z}\, e^{c_o \sqrt{\frac{\beta v_o}{V}}}; \psi(\vec{r})) \rangle_{av\,\psi} \rangle_{av\, c_o} \quad (44)$$

The expansion of the logarithm of the grand canonical partition functional as a functional Taylor series as given by Lebowitz and Percus[28] is

$$\ln \mathcal{Z}_h(y;\psi(\vec{r})) = V\beta P_h(y) + A(y;\psi(\vec{r})) \tag{45}$$

where

$$A(y;\psi(\vec{r})) = \sum_{n=1}^{\infty} \frac{1}{n!} \int_V \cdots \int d^D r_1 \cdots d^D r_n \, \mu_n(\vec{r}_1 \cdots \vec{r}_n | y) \prod_{j=1}^{n} \psi(\vec{r}_j) \tag{46}$$

and $P_h(y)$ and $\mu_n(\vec{r}_1 \cdots \vec{r}_n | y)$ are the pressure and the modified Ursell functions, respectively, of the reference system at fugacity y. One has thus

$$\mathcal{Z}(z) = \int_{-\infty}^{\infty} \frac{dc_o}{\sqrt{2\pi}} \, e^{-\frac{c_o^2}{2} + V\beta P_h(y)} \langle e^{A(y;\psi(\vec{r}))} \rangle_{av\psi}$$

$$= \sqrt{\frac{V}{2\pi\beta\upsilon_o}} \int_o^{\infty} \frac{dy}{y} \, e^{V[\beta P_h(y) - \frac{(\ln y - \ln \tilde{z})^2}{2\beta\upsilon_o} + h(y)]} \tag{47}$$

where

$$h(y) \equiv V^{-1} \ln \langle \exp A(y;\psi) \rangle_{av\psi} . \tag{48}$$

Strictly speaking, $P_h(y)$ and $h(y)$ still depend on V. If they approach their limiting value for infinite V sufficiently fast, one obtains the pressure $P(z)$ from

$$\beta P(z) \equiv \lim_{V\to\infty} \frac{1}{V} \ln \mathcal{Z}(z) = \text{l.u.b.} \{ \lim_{V\to\infty} [\beta P_h(y) - \frac{(\ln y - \ln \tilde{z})^2}{2\beta\upsilon_o} + h(y)] \} \tag{49}$$

This equation implies the expression for the density and a corrected Maxwell construction. Neglecting $h(y)$ reproduces of course the van der Waals-Maxwell equation of state. If one takes only the first and second terms of the expansion of $A(y;\psi(\vec{r}))$ in Eq. (46), the average in Eq. (48) can be evaluated exactly, and one obtains the terms through order γ^{2D-1} given by Zittartz[23] which include the term of order γ^{D} of Hemmer.[26] The further expansion of $h(y)$ was carried through by Jalickee, Siegert, and Vezzetti.[25] It is facilitated by a transformation to Gaussian random functions $\tilde{\psi}$ with a different covariance $\beta\tilde{\upsilon}$, where $\tilde{\upsilon}$ is related to the original potential υ in a similar way as the Debye shielded potential to the Coulomb potential but is of longer range than υ. There is some freedom in choosing this transformation and for a special choice it serves to extract the first correction term as a factor from the partition function. The complete expansion for the pressure, after the first correction term, is expressed as a series of connected diagrams in which hypervertices represent μ_n's and bonds the (neutralized) transformed potential $\tilde{\upsilon}$. The lowest power of γ to which any diagram can contribute is easily obtained from the number of bonds and hypervertices. The diagrams needed through order γ^{3D-1} are, for instance, the hypervertex representing μ_4 with two re-entrant bonds, and two hypervertices, each representing μ_3 with three connecting bonds. These diagrams contain the term of order γ^{2D} given by Lebowitz, Stell, and Baer.[27]

The corresponding γ-expansion of the pair distribution function was carried out by Jalickee, Siegert, and Vezzetti.[29] The pair distribution function obtained in this way vanishes, as it should, in the hard core region, if the repulsive potential has a hard core. This removes an objection against the γ-expansion stated by a previous author. This objection would have been very serious, because it would have meant that the expansion fails even for values of temperature and fugacity bounded away from the coexistence curve.

The expansion of $h(y)$ fails in an obvious manner at the van der Waals critical temperature (T_o) and fugacity, and it fails for an interval of values of y at temperatures below T_o. The values of y obtained by the Maxwell construction remain outside of this interval. The expansion is also known to fail, in a less obvious manner, for the one-dimensional model treated exactly by Kac, Uhlenbeck, and Hemmer, since it predicts a phase transition for the case $\gamma > 0$, while the model has actually, instead of the phase transition, for small $\gamma > 0$, a very high compressibility in the region near the van der Waals-Maxwell transition. Since the density fluctuations are then very large, one can understand that the expansion in the spatially inhomogeneous part of the random field fails. Conversely, it is physically reasonable to expect that the expansion yields correct results for models which have a first-order phase transition, and no abnormally large compressibility, for small γ at any fixed temperature above or below T_o in the liquid and gas region.

The obvious failure of the expansion when the temperature approaches T_o from above can be avoided by using the freedom of choice of the transformation from ψ to $\tilde{\psi}$, to obtain a modification[30] similar to the modification of the γ-expansion for the Ising model.

I would like to thank Professor C. J. Thompson for valuable comments.

REFERENCES

1. M. Kac, in Applied Probability, (L. A. MacColl, ed.), McGraw-Hill, New York, 1957.

2. L. Onsager, Phys. Rev., 65, 117 (1944).

3. L. van Hove, Physica, 15, 951 (1949).

4. D. Ruelle, Statistical Mechanics Rigorous Results, Benjamin, New York, 1969).

5. G. S. Rushbrook and H. D. Ursell., Proc. Cam. Phil. Soc., 44, 263 (1948).

6. L. van Hove, Physica, 16, 137 (1950).

7. R. Brout, Phys. Rev., 118, 1009 (1960).

8. G. A. Baker, Phys. Rev., 122, 1477 (1961).

9. M. Kac, in Studies in Mathematical Analysis and Related Topics, (Gilbarg et al., eds.), Stanford Univ. Press, Stanford, Calif., 1962.

10. M. Kac, G. E. Uhlenbeck, and P. C. Hemmer, J. Math. Phys., 4, 216 (1963); 4, 229 (1963); 5, 60 (1964); see also 5, 75 (1964).

11. N. G. van Kampen, Phys. Rev., 135, 362 (1964).

12. J. L. Lebowitz and O. Penrose, J. Math. Phys., 5, 75 (1966).

13. M. Kac and E. Helfand, J. Math. Phys., 4, 1078 (1963).

14. M. Kac and C. J. Thompson, J. Math. Phys., 10, 1373 (1969).

15. M. Kac, Brandeis Lectures, 1966, Vol. I, Gordon and Breach, New York, (1968).

16. A. J. F. Siegert, to be published.

17. B. Mühlschlegel and H. Zittartz, Ann. d. Physik, 175, 553 (1963).

18. A. J. F. Siegert and D. J. Vezzetti, J. Math. Phys., 9, 2173 (1968).

19. F. Dyson, Commun. Math. Phys., 12, 91, 212 (1969).

20. G. Horwitz and H. B. Callen, Phys. Rev., 124, 1757 (1961).

21. C. J. Thompson, A. J. F. Siegert and D. J. Vezzetti, J. Math. Phys., 11, 1018 (1970).

22. R. B. Griffiths, J. Math. Phys., 8, 478 (1967).

23. J. Zittartz, Z. Physik, 180, 219 (1964).

24. J. B. Jalickee, A. J. F. Siegert, and D. J. Vezzetti, J. Math. Phys., 10, 1442 (1969).

25. M. Coopersmith and R. Brout, Phys. Rev., 130, 2539 (1963).

26. P. C. Hemmer, J. Math. Phys., 5, 75 (1964).

27. J. L. Lebowitz, G. Stell, and S. Baer, J. Math. Phys., 6, 1282 (1965).

28. J. L. Lebowitz and J. K. Percus, J. Math. Phys., 4, 1495 (1963).

29. J. B. Jalickee, A. J. F. Siegert, and D. J. Vezzetti, J. Math. Phys., to be published.

30. See ref. 24, p. 1451.

A REVIEW OF SUPERFLUIDS AND SUPERCONDUCTORS

Paul C. Martin

Harvard University
Cambridge, Massachusetts

Just about three years ago, in Copenhagen, I spoke on recent theoretical advances in the understanding of superfluid helium. I did so in a setting which I think is very similar to the present conference. Our meeting at Northwestern is formally dedicated to Professor Uhlenbeck, but anyone who has regularly attended the biennial IUPAP meetings on statistical mechanics will tell you that they have been so influenced by his interests, ideas, and criticisms, that it is hard to think of those meetings in a different light than the present one.

I was therefore reluctant to speak about superfluids at this conference for three reasons: first, because I have not added to the subject since I last reviewed it in a similar context; second, because I believe that the conventional statistical mechanics of superfluids and superconductors are basically, and even semiquantitatively understood; even if they are not understood at the level of rigor some here may wish for; and thirdly, because I understand that Professor Uhlenbeck has worked on these problems during the past two years, and I am wary that he will raise some argument which will undermine this complacency.

At any rate, the understanding of superfluidity does represent an achievement of the past ten to fifteen years, and so it is appropriate to review it at a meeting such as this one. Furthermore, superconductors and superfluids are likely, in the future, to provide a testing ground for theoretical and experimental study of two of the most fundamental problems in statistical mechanics: the phase transition and the metastable state. I shall therefore divide my talk into two parts: a summary of the generally accepted and acceptable basis for describing the thermodynamics and hydrodynamics of the superfluid, and a brief resume of the currently interesting nonequilibrium problems associated with vortex motion, supercurrent decay, and fluctuations.

Let me begin by distinguishing superfluidity and superconductivity from the other properties of superfluid substances. In helium, typical phonon energies and even the binding energies are substantially larger than the transition temperature. For this reason, most of the low temperature thermodynamic properties of helium make virtually no reference to the existence of superfluidity. This is the case, for example, with the recent impressively accurate calculations of Feenberg and a number of his students and their collaborators. It is also the case with a number of calculations that deal with the temperature dependent properties of phonons and rotons. The former variational perturbative calculations give quantitative values for the thermodynamic properties of helium and its elastic constants. The latter give satisfactory estimates for ultrasonic attenuation and the temperature dependence of the sound velocity, in terms of an interacting excitation theory which employs a phenomenological dispersion law and phenomenological elastic constants but makes no reference to superfluidity. Superfluidity in helium has a profound and dramatic effect on relatively few of the thermally excited modes and very little of the correlational energy. In superconductors, the transition affects the excitation in a more striking way: the thermally excited electrons are usually modified by the appearance of an energy gap. But even here, the correlational energy change is small.

I. BASIS FOR SUPERFLUIDITY: LONG-RANGE ORDER

It is now generally understood that the crucial aspect of the superconducting or superfluid state is not the nature of the excitation spectrum but the occurrence of a macroscopically occupied or coherent mode. The most general state which will exhibit superfluid properties is difficult to specify. Indeed, since real superconductors have finite extent, the physical question of infinite conductivity is different from the mathematical one. However, a sufficient condition for superfluidity, in an infinite system, the one due to Onsager and Penrose, is that the one particle density matrix of the superfluid

$$N_\Omega \rho_\Omega(\vec{r}\,\vec{r}\,') = \langle \psi^\dagger(\vec{r})\psi(\vec{r}\,')\rangle_\Omega$$

$$= N \sum_s \omega_s \int \Psi_s^*(\vec{r}\vec{r}_2 \ldots \vec{r}_{n_\Omega})\Psi_s(\vec{r}\,'\vec{r}_2 \ldots \vec{r}_{n_\Omega})d\vec{r}_2 \ldots d\vec{r}_{n_\Omega}$$

has a spectral decomposition

$$\langle \psi^\dagger(\vec{r})\psi(\vec{r}\,')\rangle = n_{\alpha_c}\varphi^*(\vec{r})\varphi_{\alpha_c}(\vec{r}\,') + \int d\alpha\, n'(\alpha)\varphi_\alpha{}^*(\vec{r})\varphi_\alpha(\vec{r}\,')$$

in which a discrete eigenvector, φ_{α_c}, called the condensate, has finite weight, n_{α_c}, as the volume, Ω, increases. In this expression, the functions $\varphi_{\alpha}(\vec{r})$ have been normalized in the form

$$\frac{1}{\Omega} \int_{\Omega} |\varphi_{\alpha}(\vec{r})|^{2} d\vec{r} = 1$$

so that the trace of the density matrix divided by the volume, that is, the density, is given by

$$n = n_{\alpha_c} + \int d\alpha \; n'(\alpha) \quad ,$$

where n_{α_c} and $n'(\alpha)\, d\alpha$ are positive. It is conventional to call n_{α_c} the condensate density and the integral, the depletion. In a perfect Bose gas at zero temperature, there is no depletion.

It may well be sufficient for a very small fraction of the modes with a sufficiently singular density to exhibit effects similar to those produced by a condensate of the type just postulated. In fact, in one or two dimensions, this is the only possibility: it can be shown that the translationally invariant equilibrium state has no condensed mode or long-range order. We shall return to this point later.

For a superconductor, it is:

$$\psi_{pair}(\vec{r}) = \int \psi_{\uparrow}(\vec{r} + \frac{\vec{s}}{2})\, \psi_{\downarrow}(r - \frac{\vec{s}}{2})\, \varphi_{pair}(\vec{s})\, ds$$

which describes fermi pairs and has the properties just outlined. Because the internal structure of the pair complicates the discussion somewhat, I shall give more weight to helium and less to superconductors than one ought to.

To say that the description I have given is the proper one for a superfluid is to say that as a result of interparticle interactions and statistics, the equilibrium states of the superfluid have this form. It can be argued that this long range order is maintained in three dimensions by the same perturbation theory which shows that the normal state is unstable. It is also possible to use a sophisticated perturbation theory which shows that long-range order is not produced in less than three dimensions, but this is not the one conventionally employed.

In addition, except for the anomalous free boson gas, it can be shown that the free energy of states with more than one condensed mode or with no condensed mode is higher. None of these arguments constitute a rigorous proof that the equilibrium density matrix of superfluid helium has the alleged form, but there appears to be no appealing, let alone compelling theoretical or experimental argument for believing the picture is not correct in three dimensions.

II. THE CONDENSATE AND SUPERFLUID VELOCITY

An equivalent description of superfluidity can be achieved in terms of partially coherent states, states in which the number operator is not diagonal and in which as a result of coherence

$$<\psi(\vec{r})> = \sqrt{n_c(\vec{r})}\, e^{i\varphi_c(\vec{r})} = \varphi_{\alpha_c}(\vec{r})\sqrt{n_{\alpha_c}}$$

that is,

$$<\psi^{\dagger}(\vec{r})\,\psi(\vec{r}\,')> = <\psi^{\dagger}(\vec{r})>\ <\psi(\vec{r}\,')>$$

$$= \sqrt{n_c(\vec{r})}\, e^{-i\varphi_c(\vec{r})}\sqrt{n_c(\vec{r}\,')}\, e^{i\varphi_c(\vec{r}\,')} \ .$$

The phase plays the role of a velocity potential for the superfluid. Its absolute value has no significance but differences in potential do. The gradient of the potential gives the superfluid momentum flow

$$\vec{p}_s(\vec{r}) = \hbar\vec{\nabla}\varphi_c(\vec{r}) \ .$$

The single valuedness of the condensate phase, φ_c, in multiply connected regions implies vorticity and flux quantization. Vorticity quantization has been strikingly demonstrated in the experiments of Rayfield and Reif in which $\hbar$

has been measured to an accuracy of one percent. In superconductors where $\frac{e}{m}(\vec{p}_s(\vec{r}) - \frac{2e}{c}A(\vec{r}))$ is the velocity of the supercurrent, the quantization of $\vec{p}_s(\vec{r})$ and the vanishing of the current inside the superconductor imply flux quantization in units of $hc/2e$. The consequences of this quantization condition and the equilibrium time variation of the phase that follows from Galilean invariance have been even more strikingly verified. As a result of measurements of the time variation of a phase difference, the fine structure constant has been determined to be $\alpha^{-1} \cong 137.060$.

In a uniform system at low temperatures, the phase of the condensate wave function must be $\varphi_c(\vec{r}) = \vec{p}_s \cdot \vec{r}/h$, so that $\vec{p}_s$ is the constant superfluid momentum. The amplitude of the condensate wave function $\sqrt{n_{\vec{p}_s}}$ is also constant and the momentum distribution is

$$n(\vec{p}) = n_{\vec{p}_s}\, \delta(\vec{p} - \vec{p}_s) + n'(\vec{p}).$$

$n(\vec{p})$ varies with $\vec{p}_s$, and there appears to be no metastable solution, i.e., no state with long lifetime if $\vec{p}_s$ is greater than some critical value.

The only truly equilibrium state of the physical system is the one which has $\vec{p}_s = 0$. (Mathematically, the states with $p_s = 0$ and different values of the absolute phase must be distinguished. Physically, we may also distinguish them in principle with a test coupling but not in a closed uniform system.) Translationally invariant states with $p_s \neq 0$ have higher free energy. However, the excitations in uniform states with $p_s \neq 0$ are almost orthogonal to the excitations with $p_s = 0$ when the volume goes to infinity. As a result states with $p_s \neq 0$ are metastable in this limit. The processes which produce changes in p_s are qualitatively different from those which bring normal currents to rest. They require an understanding of nontranslationally invariant states with vortices and their overlap with the translationally invariant states.

III. DISTINCTION BETWEEN n_o/n AND n_s/n

As I indicated, when a mode is macroscopically occupied with well-defined phase and amplitude, it is called coherent. The nature of the particular partially coherent states of helium and superconductors is determined by the interaction and the temperature. They are more complicated than the coherent states of an oscillator produced by a macroscopic external force, in which there is a Poisson distribution of specific width for the coherent amplitude. The distribution of amplitudes of the condensate in helium and superconductors, enforced by its interaction with the remainder of the many body system, is also sharply peaked about n_o, but it must be calculated, and it depends on details of the microscopic interaction. Although calculations that demonstrate the distinction have existed for some time, there seems to be a persistent tendency to let the hope that the superfluid were as simple as a fully coherent noninteracting photon state give rise to the claim that it is. The claim is unfortunately incorrect.

Much of the desire for the overly simple picture of helium stems from the hope of replacing the well defined, microscopic parameter, n_o/n, the fraction of particles in the condensed mode, by the well-defined thermodynamic parameter n_s/n, the superfluid fraction. The thermodynamic parameter n_s/n behaves in an appealing way and can be simply measured by a variety of methods: the velocity of second sound, the moment of inertia of helium in an appropriately chosen container, or by experiments involving counterflow. No one would try to identify n_s with n_o if there were incontrovertible experimental evidence to the contrary. However, while the existence of

n_o plays a fundamental role in the microscopic explanation of superfluidity, n_o does not lend itself to simple direct measurement. It may just possibly be determined by measuring the momentum distribution in helium with high energy neutrons, but the determination of the momentum distribution of equilibrium superfluid helium from these experiments is quite difficult. A more accurate calculation of the remainder of the helium correlation function than we presently have is required. The recent Chalk River experiments are quoted as being best fit by taking $n_o/n \sim 20\%$ in contrast with theoretical estimates of $n_o/n \sim 10\%$, but experimentally the errors are still huge enough to allow any value of n_o/n.

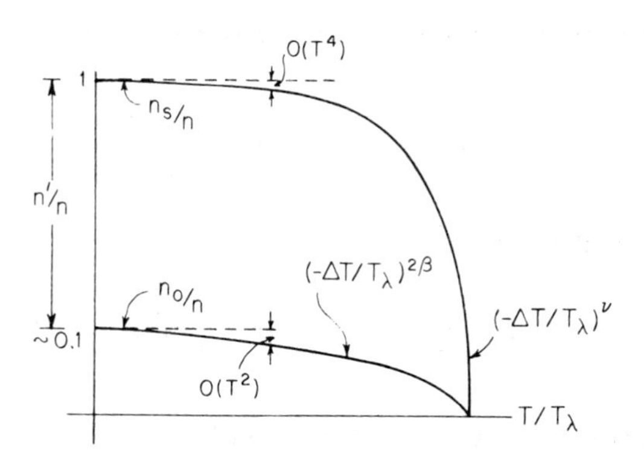

FIG. 1. n_o and n_s vs. temperature.

Perhaps the best way of distinguishing what we believe to be the difference between n_o and n_s is by examining the theoretical predictions. The upper curve for n_s (Fig. 1) has been verified experimentally in the three indicated respects. The lower curve has only been tested by the marginal neutron experiment at low temperatures. It has been shown that in perturbation theory $n_s \geq n_o$ so that $n_s \neq 0$ if $n_o \neq 0$, and also at T=0, $n_s = n$.

IV. DYNAMICS AT A MACROSCOPIC LEVEL

It is worth stressing that n_s is only defined in a meaningful fashion for uniform systems or systems which vary slowly in time and slowly in space compared to mean free paths and correlation lengths. Thus the microscopic equations we could introduce for the particle density

$$\delta_{ij}\, n(\vec{r}) = n_{s_{ij}}(\vec{r}) + n_{n_{ij}}(\vec{r})$$

and the current density

$$j_i(\vec{r}) - n(\vec{r})\,\vec{v}_{s_i}(\vec{r}) \equiv n_{n_{ij}}(\vec{r})(\vec{v}_{n_j}(\vec{r}) - \vec{v}_{s_j}(\vec{r}))$$

must be augmented by a macroscopically defined quantity $\vec{v}_n$ in order to have content. Any of the conventional definitions, for example,

$$\left.\frac{\partial \epsilon}{\partial \vec{j}_i}\right)_{s,\vec{v}_s} = m\vec{v}_{n_i}$$

where ϵ is the energy and s the entropy, can be shown to imply the remainder of the two fluid description in a nearly uniform system.

With the restriction to uniform equilibrium states, a convenient way to define the superfluid density is by the rotating bucket experiment or in superconductors, by the Meissner effect. The retarded transverse current commutator which describes this experiment provides one convenient definition for the computation of the normal fluid density $n_n = n - n_s$. Of necessity, the derivation of the two fluid model is restricted to infinitesimal disturbances since finite amplitude disturbances could lead to vortex production. To derive a hydrodynamic theory which is appropriate when vortices are present, it is necessary to average over many of them. As you may recall, these vortices tend to align themselves in specified lattice configurations when fields or rotations are applied. The resultant hydrodynamic theories are therefore extremely complicated.

Even when there are no vortices, the corresponding hydrodynamical theory for superconductors is of little interest. Electromagnetic effects prevent the supercurrent from ever oscillating slowly in time and space. The two fluid models would hold for the hypothesized low temperature phase of He^3 in which pairs condense. Of the two fluid parameters, the superfluid velocity, alone, has a more microscopic identification. Along with the magnitude of the condensate, it serves to characterize the nature of the microscopic vortex.

To understand the dynamics of a vortex or collection of vortices, it is necessary to have some notion of their microscopic structure and possible forms of movement. In a superconductor where the vortex core cross section extends over perhaps 10^5 atoms, this problem is particularly significant. For many purposes, and in many limits, it seems that the Landau-Ginsburg equation suffices. However, it also seems likely that there is no macroscopic equation for the interacting fluid with vortices that describes the vortex motion with the same degree of rigor with which hydrodynamics describes small oscillations of the vortex-free system. Instead there are equations of greater complexity and statistical character analagous to those for Brownian motion. They must include nondissipative terms which represent vortex interaction, internal degrees of freedom associated with vortex oscillations, and damping terms which describe losses in the core and losses resulting from collisions with elementary excitations.

V. MICROSCOPIC DYNAMICS

For the most part, we have a good idea about the elementary excitations of the superconductor and superfluid. These do vary substantially from one superconductor to the next. For example, some superconductors do not have an energy gap. In helium we know about the phonons and rotons. Near the phase transition there are many complications. Away from the phase transition there are few. One of these is the nature of the dynamic fluctuations of the condensate amplitude. It is difficult to calculate them reliably because the phase fluctuations are so large and hard to separate. It is difficult to measure them for the same reason that the momentum distribution is difficult to measure.

To be more precise, we know that at small wavenumbers, the positive definite spectral function

$$\omega^{-1}A(k\omega) = \omega^{-1}\int <[\Psi(\vec{r}t),\ \Psi^{\dagger}(\vec{r}'t')]> \exp[ik\cdot(\vec{r}-\vec{r}')-i\omega(t-t')]$$

has a dominant part due to the phonons, that behaves like

$$\omega^{-1} A(\vec{k}\omega) = \frac{2\pi n_o}{m n_s c^2 k^2} [\delta(\omega - ck) + \delta(\omega + ck)] \ .$$

This contribution does not exhaust the frequency sum rule, and the dominant frequencies in the remainder which bear on amplitude fluctuations have not been reliably determined.

VI. THE MOMENTUM DISTRIBUTION AND PHASE FLUCTUATIONS

For small momenta the dominating phase fluctuations give rise to a momentum distribution which reflects the fact that the long wavelength superfluid fluctuations are essentially classical. In particular, the answer may be understood to result from having a mean energy

$$n_s(T) \frac{\langle p_{sk}^2 \rangle}{2m} \cong \frac{1}{2} k_B T$$

in each long wavelength mode of the superfluid momentum. Since these fluctuations represent phase fluctuations, and since the momentum distribution is related to phase fluctuations because quantum field fluctuations determine both

$$p^2 n(\vec{p}) = \langle \psi^\dagger_{\vec{p}} p^2 \psi_{\vec{p}} \rangle \cong \frac{n_o(T)}{h^3} \langle (\hbar k \varphi_{\vec{k}})^2 \rangle = \frac{n_o(T)}{h^3} \langle p_{s\vec{k}}^2 \rangle$$

we have

$$n(p) = n_o(T)\left[\delta(p) + \frac{k_B Tm}{h^3 p^2 n_s(T)} + \frac{mc}{h^3 np}\right].$$

In this last formula we have also included the contribution of the condensed mode and the zero-point motion contribution that comes from a real derivation. This formula is based on perturbation theory. Rigorously, one also can prove that it is an inequality with the n_s's in the denominator replaced by n's . The zero-point motion reflects the interaction through the appearance of the sound velocity c . In addition, although it cannot be seen so readily, the second term has a coefficient

$$(2 - \frac{c\Omega}{c+\Omega})$$

where Ω is of the order of the size of the system. For any interaction this expression equals unity, but for no interaction it equals 2. The free bose gas has a momentum distribution

$$n(p) = n_o(T)\left[\delta(p) + \frac{2k_B Tm}{h^3 p^2 n_o(T)}\right].$$

The peculiarity of the free gas has been pointed out long ago. It gives rise to certain ambiguities and arguments about whether the free gas is really coherent after all. Since any weak interaction eliminates these amplitude fluctuations, the ambiguities are not relevant to physics. While a weak interaction eliminates the large amplitude fluctuations that correspond to large number fluctuations in the free bose gas, it leaves the phase fluctuations unaltered. These fluctuations, whose existence follows from the phase degeneracy of equilibrium states, or, if you prefer, the Goldstone theorem, are connected with several important properties of helium.

In the first place, they are too large to allow a phase transition to a state with a condensate in one or two dimensions at finite temperature, and in one dimension at $T = 0$, since in each case condensation would imply a momentum distribution which is inadmissible since $\int d\vec{p}\, n(\vec{p})$ diverges.

In the second place, they permit us to infer some important properties of the phase correlations. Our expression

$$\int d\vec{r} e^{-i\vec{k}\cdot\vec{r}} \frac{1}{2}\langle(\varphi(\vec{r}) - \varphi(0))^2\rangle \sim \varphi^2_{\vec{k}} \sim \frac{mk_B T}{\hbar^2 n_s(T) k^2}$$

suggests that in different dimensions the phase fluctuations are given by

$$\langle e^{i\varphi(r)} e^{-i\varphi(r')}\rangle \cong \exp \int \frac{d^d k}{(2\pi)^d} e^{i\vec{k}\cdot(\vec{r}-\vec{r}')} \frac{mk_B T}{\hbar^2 n_s(T) k^2}$$

$$\cong \exp\left[-\frac{mk_BT}{\hbar^2 n_s(T)}\,|\vec{r}-\vec{r}\,'|\right] \qquad d=1$$

$$\cong (|\vec{r}-\vec{r}\,'|)^{-(mk_BT/\hbar^2 n_s(T)2\pi)} \qquad d=2$$

$$\cong e^{i(\varphi(r)-\varphi(r'))}\exp\left[-\frac{mk_BT}{4\pi\hbar^2 n_s(T)}\,\frac{1}{|\vec{r}-\vec{r}\,'|}\right] \qquad d=3$$

These formulas which are purely inferential suggest that in two dimensions there may be a phase transition without long-range order because at and below some temperature the exponent of $|\vec{r}-\vec{r}\,'|$ gives rise to a nonintegrable correlation function.

VII. PHASE TRANSITION

We may also infer an important property of the actual three dimensional phase transition. Near the transition temperature, the coherence length ξ determines the volume over which the fluctuations occur. Since that volume is equal to both $\xi^{d-1}/n_s(T)$ and ξ^d in three or more dimensions we have

$$n_s(T) \sim \frac{1}{\xi} \sim \frac{(T_c - T)^{\nu}}{T_c} \quad .$$

In three dimensions, laws predict $\nu d = 2-\alpha$ where α is the exponent that describes the specific heat singularity. The consequent prediction of $\nu = 2/3$ is well borne out experimentally. In three dimensions this is not very different from the value expected for the singularity of $n_0(T)$. Likewise, molecular field theory would give a linear behavior for both.

If scaling laws are true, an understanding of the easily measurable properties of helium would be complete if the logarithmic specific heat were well understood. At the moment, I do not understand it although some others claim to. I shall say nothing about the very interesting dynamical critical phenomena which have been observed and partially explained in helium since I believe Hohenberg will discuss them.

VIII. METASTABILITY BELOW T_c

We have argued that long-range order is absent in two dimensions because the fluctuations in phase do not die off sufficiently rapidly with distance. While these phase fluctuations rule out a phase transition to a condensed equilibrium state, it is not clear whether they rule out a transition. In any event they do not give rise to resistance. The superconducting phase may be likened to the turning of a long spring. A small uncertainty in the spring pitch will prevent us from knowing how many turns occur over a very long distance, but the absence of this definite information about the phase is not indicative of a loss mechanism. Such a system may be best described at low temperatures as an ensemble of coherent states in two or even in one dimension despite the uncertainty in phase difference between distant points. When this is the proper description, there will be a qualitative change in resistance at temperatures not very different from the mean field transition temperature. At lower temperatures where coherence extends over many particles, only a relaxation of a current will occur because coherence is broken.

To be concrete, let me recall that in a superconductor the correlations responsible for the transition typically extends thousands of angstroms. In working with extremely thin wires and whiskers, one effectively eliminates the long-range correlations in two of the three dimensions. In the resultant effectively one dimensional geometry, a decrease in the supercurrent by one quantum is tantamount to cutting one notch out of the circular spring, (in contrast to the wavering of the spring which eliminates long-range order). Calculating the lifetime of such a supercurrent depends on being able to calculate the rate at which breaks in the spring, that is, normal regions the size of a vortex core occur in the superconductor. Physically these breaks correspond to the passage of flux quanta or vortices through the ring. We would therefore like to calculate from first principles a kind of frequency of macroscopic fluctuations, a frequency which we expect will include what chemists call an activation energy.

Phenomological calculations of the lifetime of supercurrents have been carried out on the basis of this picture by Langer, Fisher, Ambegaokar, Halperin, and McCumber. Using the basic picture that the slowing of the supercurrent requires a fluctuation in which the condensate density vanishes, since only then can the phase change, they calculate the likelihood of these fluctuations using a semimicroscopic, i.e., Landau-Ginsburg equation approach.

Their calculations appear to show that the coherent state of a one dimensional superconducting sample has a very small, but nonvanishing resistance below a characteristic temperature of the order of the three dimensional T_c. Specifically, the ratio of the resistance to the normal state resistance found by Halperin and McCumber has the dominant factors

$$\frac{R}{R_n} \cong \left(\frac{\Delta T}{\Delta T_o}\right)^{9/4} \left(\frac{T_c}{T}\right)^{3/2} \exp\left[-\left(\frac{\Delta T}{\Delta T_o}\right)^{3/2}\left(\frac{T_c}{T}\right)\right]$$

where the width of the transition ΔT_o varies inversely as the diameter of the wire to the 4/3 power. The width is of the order of 1 millidegree for a tin whisker $1/10\ \mu$ in diameter. It takes a very long wire at low temperature to have a measurable supercurrent resistance.

In two dimensions where long range order can almost occur, calculations suggest that the conductivity might be inversely proportional to a power of the current and infinite for vanishing current. We picture this resistance as a consequence of the formation and separation of vortex pairs. In three dimensions it seems likely conductivity is infinite for vanishing current and that the rate at which currents decay is incredibly small, of the order of $\exp(-J_c/J)$ when $J \ll J_c$, the critical current. Thus, below T_c in superconducting systems theorists are beginning to seriously study questions bearing on metastable as well as stable states; they are doing so by investigating how the metastability comes about, calculating the rate at which the spontaneous fluctuations which generate and destroy vortex lines occur, and the rate at which these spontaneous fluctuations can escape from calculated local free energy minima through calculated potential barriers. Presumably at the phase transition these barriers vanish and vortex fluctuations grow uninhibitedly.

IX. FLUCTUATIONS ABOVE T_c

While they are somewhat less novel, the properties of the superfluid and superconductor right above the transition are also far from fully understood. In particular, the behavior of the incipient supercurrent in a superconductor and its diamagnetic and conduction properties have received a great deal of attention in the last couple of years. The diamagnetism is relatively well understood, but the resistance above T_c seems to be in a confused state. A simple theory, based on the diffusion of fluctuating pairs, according to the time dependent Landau-Ginsburg theory, was first derived on the basis of more microscopic arguments by Larkin and Aslamosov. The theory is appealing but seems to be incomplete. Attempts to complete it produce further difficulties. Whether they are due to our ineptness in treating the mathematical complexities of the problem, or whether they indicate a failure of the Landau-Ginsburg fluctuation theory for transport properties even far from the transition temperature remains to be settled.

In short, twenty years ago superfluids were a real mystery; ten years ago they represented an active and exciting area, with new phenomena being discovered at frequent intervals. Now superconductors are so well understood that more subtle questions than those normally studied in statistical mechanics are being considered. They are also being extensively studied experimentally for two related reasons: superconductors are practically important and superconductors are amenable to quantitative study because their properties can be widely varied. We may hope that the extensive experimental work and related theory will help us in understanding evaporation and other nucleation processes. But that remains to be seen.

At any rate, it is in this direction, and the area of the phase transition, that superconductivity and superfluidity research has turned, and it is in this area that we may at least hope for important new insights.

DYNAMIC PHENOMENA NEAR A CRITICAL POINT

P. C. Hohenberg

Bell Telephone Laboratories Incorporated
Murray Hill, New Jersey

I. INTRODUCTION

As we heard from Professor Domb, there has now accumulated a substantial body of information concerning the thermodynamic singularities and long-range spatial correlations which exist in systems near critical points. In contrast to these equilibrium or static properties, the dynamic behavior of such systems has until very recently only been investigated much more qualitatively. In the past two or three years, however, a number of important advances have been made, which have provided some of the most detailed and precise experimental tests of our understanding of critical phenomena. What I have been asked to do, in the present talk, is to review some of these developments, and to assess the present status of our understanding of critical dynamics. I must apologize at the outset, since this review will of necessity be sketchy, and biased by my own interests and point of view.[1]

Let me begin by summarizing very briefly the history of <u>static</u> critical phenomena, as it developed over the past few decades. First, there was a phenomenological theory, that of Van der Waals, or somewhat more generally, the universal theory of second-order phase transitions of Landau.[2] This theory gave a very good qualitative account of the phenomenon of the critical point, and it provided a qualitatively correct description of most static properties. The theory was universal, in that it predicted essentially the same behavior for all systems near the critical point. It was universal but, as you all know, it was not correct.

This we learned from the second item in our history of static critical phenomena, namely, the existence of a <u>nontrivial soluble model</u>, the two-dimensional Ising model,[3] whose behavior violates the predictions of the Landau theory. In particular the specific heat at constant magnetization diverges in the exact theory, whereas it remains finite in the approximate phenomenological theory.

The third important point, which demonstrated the practical relevance of exact theories, was the experimental measurement on real systems, of thermodynamic quantities. Most important, I believe, was the measurement of the specific heat of He^4 near the λ point,[4] and of Argon[5] near its critical point. Those experiments showed not only that the Landau theory was incorrect, but they also displayed a temperature dependence which was strikingly similar to the logarithmic divergence found by Onsager for the two-dimensional Ising model.

Fourthly, let me mention the "numerical experiments" which we heard about in some detail from Professor Domb, based on computer evaluations of series expansions in various model systems.[6] This work again showed the relevance of precise calculations, since many of the singularities found in the models were very similar to those inferred from experiments on real systems.[7]

The fifth and very important item is one which I would like to call hard theory. Now hard is the opposite of soft, so by hard theory I will mean that collection of theorems about phase transitions which we know firmly to be true. Clearly, item two above belongs to this category. But unfortunately, hard is also the opposite of easy, i. e., hard theory is difficult, so that the number of relevant results is not as large as one might hope. Nevertheless, we are beginning to accumulate a body of hard theory on static critical phenomena, such as various inequalities, analyticity theorems, statements about the existence of phase transitions, etc.[6] The numerical work in item four above, for instance, does not belong in that category. In a sense this theory is both soft and hard: it is soft because it only yields information about the critical point by an extropolation which is not mathematically rigorous. On the other hand, it clearly is hard, in the sense that a lot of very difficult and clever mathematical physics goes into carrying out the series extrapolations!

In the absence of a definitive and complete body of hard theory of the critical point, it has been fruitful to resort to a sixth ingredient, namely, the development of _alternate phenomenological theories_ that would replace the universal theories of van der Waals and Landau. The theory which seems to have gained the most popularity in recent years is the so-called similarity or scaling theory,[8,9] which has at least helped one to correlate and compare experiments on many different systems, and may even be exactly correct for some realistic models.

With this brief outline of the development of static critical phenomena, let us see where we stand concerning dynamic properties. Firstly, corresponding to the classical phenomenological Landau theory, there is the Van Hove theory[10] of critical slowing down. This theory is not nearly so complete or universal as the Landau theory, since dynamical properties depend much more sensitively on the details of the Hamiltonian. Nevertheless, there exists a universal concept, if not a universal theory.

The second ingredient which was so crucial for static phenomena, the exactly soluble nontrivial model, is completely absent for dynamic phenomena.

The third item, detailed experimental investigation, was almost completely absent five years ago, but has developed remarkably rapidly since then. It is this development, along with the sixth item, namely, a phenomenological approach to dynamic critical phenomena, which will occupy the bulk of my talk. As for the fourth and fifth points, numerical experiments and hard theory, very little exists at this time, and I shall confine my discussion of these to a few brief remarks at the end.

II. CRITICAL SLOWING DOWN

Let us consider the case of a Heisenberg ferromagnet in the paramagnetic phase $(T > T_c, H = 0)$. Van Hove's argument[10] may be sketched by writing down the macroscopic equation which governs the slow variations of the magnetization in space and time,

$$\partial\vec{M}(\vec{r},t)/\partial t = D\nabla^2\vec{M}(\vec{r},t). \tag{1}$$

The spin-diffusion coefficient D satisfies the Einstein relation

$$D = \Lambda/\chi^M \tag{2}$$

in terms of the transport (Onsager) coefficient Λ, and the magnetic susceptibility χ^M. Van Hove argued that the transport coefficient Λ depends primarily on the short-range behavior of the system and should therefore remain finite at the critical point. The susceptibility, on the other hand, was known to diverge at the critical point, so that D was predicted to vanish at T_c. When the mean-field temperature dependence was used for χ^M, the diffusion constant was predicted to be linear in $T - T_c$. As Van Hove pointed out, this slowing down of the spin diffusion would show up as a reduction of the inelasticity of neutron scattering by magnetic systems, as T approached T_c. Attempts to observe this reduction were not entirely successful,[11] and it is only recently that the reasons for this failure have begun to be clarified. In particular, Marshall[12] pointed out that most neutron experiments are not done at wavelengths large compared to the correlation range near T_c, and that therefore the Van Hove theory does not apply. The alternative theory that he outlined was not very specific, but it did suggest that "some remnant of spin-wave motion should exist above T_c."

A slightly different line of investigation was initiated by Fixman,[13] who attempted to calculate various transport coefficients at the critical point of simple fluids and fluid mixtures. The main physical idea contained in Fixman's calculations is that the long-range spatial correlations predicted by the Ornstein-Zernike theory should lead to enhanced fluctuations and anomalous transport properties. The quantitative predictions which followed from this interesting idea could not be expected to be correct, since the temperature dependence of the Ornstein-Zernike correlation function is known to be wrong. Nevertheless, the notion that long-range spatial correlations can affect the transport coefficients is an interesting one. It has led Kawasaki and co-workers,[14] and also Kadanoff and Swift,[15] to develop a more general semiphenomenological approach which is less tied to the mean-field and Ornstein-Zernike theories.

From a purely phenomenological point of view, it is natural to ask whether the static scaling laws can give any information about dynamic properties, which could serve to test their validity or to broaden their range of applicability. Such an approach was followed by Ferrell and co-workers[16] in their study of the λ point of He^4. These authors predicted an anomalous damping of second sound below T_λ and a singular thermal conductivity above T_λ. Subsequently, Halperin and I reformulated and generalized[17] the work of Ferrel et al., and applied the theory to magnetic systems. I shall attempt to outline for you the physical basis for this "dynamic scaling" approach and to describe briefly some of the important predictions of the theory.

III. FORMULATION OF DYNAMIC SCALING

The most convenient language in which to describe the scaling approach to phase transitions is the language of correlation functions. Let $A(\vec{r}, t)$ be a Hermitian operator, such as the z-component of magnetization or the energy density. The dynamic and static autocorrelation functions for A are defined by

$$\hat{C}^A(\vec{r}, t) \equiv \frac{1}{2} < \{ A(\vec{r}, t) - < A(\vec{r}, t) > , A(0, 0) - < A(0, 0) > \} > ,$$

$$\equiv \int \frac{d^3k}{(2\pi)^3} \int \frac{d\omega}{2\pi} \, e^{i(\vec{k}\cdot\vec{r} - \omega t)} C^A(\vec{k}, \omega), \tag{3}$$

and

$$\hat{C}^A(\vec{r}) \quad \hat{C}^A(\vec{r}, t{=}0) \quad \int \frac{d^3k}{(2\pi)^3} \, e^{i\vec{k}\cdot\vec{r}} C^A(\vec{k}) , \tag{4}$$

where the angular brackets indicate an equilibrium expectation value, and the curly brackets an anticommutator. Similar definitions can be given for the case where A is a vector operator, such as the total magnetization $\vec{M}$ or the staggered magnetization $\vec{N}$, or a nonHermitian operator.

In the limit of long wavelengths, the correlation functions are related to macroscopic thermodynamic and hydrodynamic properties of the system, and the asymptotic form of $C^A(\vec{k}, \omega)$ as $k \to 0$, may be largely determined by these properties. On the other hand, systems near a critical point are characterized by large fluctuations in some of their variables, and correlations which extend over very large distances. It is customary to designate one particular variable, whose fluctuations are particularly large, as the order parameter ψ, and to designate its correlation length as <u>the</u> correlation length ξ. As the critical point is approached, the length ξ tends to infinity. One would therefore expect that the range of k values where the asymptotic $(k \to 0)$ form of a correlation function $C^A(\vec{k}, \omega)$ applies, would itself tend to zero as the critical point is approached. It is the central tenet of the scaling hypothesis, that ξ is the only important macroscopic (temperature dependent) length near T_c, and the behavior of the correlation functions is essentially determined by the ratio of the wavelength k^{-1} to the correlation length ξ. (We shall only be concerned with the region $k^{-1} \gg a$ and $\xi \gg a$, where a is a typical microscopic length such as the lattice spacing.)

This "scaling" behavior can best be illustrated by the graph in Fig. 1, where the wave number k is plotted on the ordinate and the inverse length ξ^{-1} on the abscissa. The origin $\xi^{-1} = 0$ is the critical point $T = T_c$, and the disordered phase is on the right $(T > T_c)$, whereas the ordered phase is on the left $(T < T_c)$. Three asymptotic regions may be identified in the (k, ξ^{-1}) plane, in each of which the correlation function $C^\Psi(\vec{k})$ has different characteristic behavior.

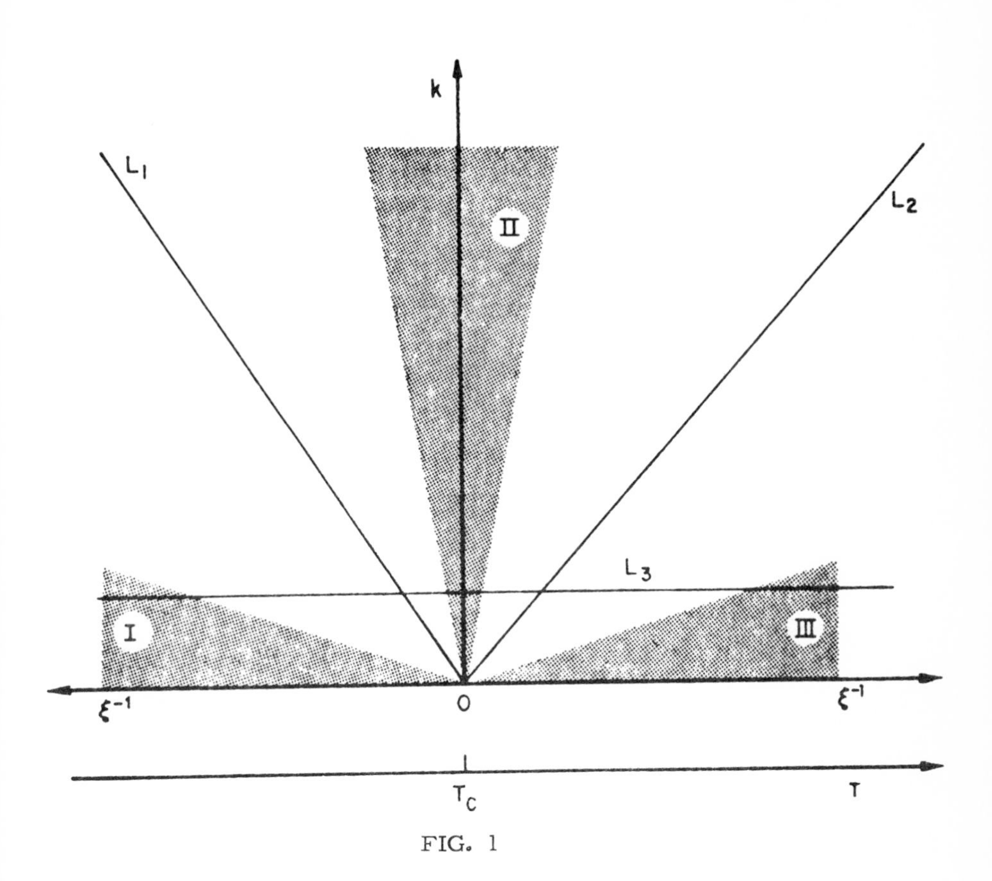

FIG. 1

The macroscopic domain of wave vector k and coherence length ξ, defined by the conditions $ka \ll 1$, $\xi/a \gg 1$. In the three shaded regions the correlation functions have different characteristic behaviors. These regions are defined by ($k\xi \ll 1$, $T < T_c$), ($k\xi \gg 1$, $T \approx T_c$), and ($k\xi \ll 1$, $T > T_c$), respectively. The asymptotic forms for these regions merge when extrapolated to the lines L_1 or L_2 ($k\xi = 1$ for $T < T_c$ and $T > T_c$, respectively). An experiment done at constant k (line L_3) will pass through all three regions as the temperature is varied. (Reprinted from Ref. 17.)

The shaded region marked I, corresponding to $k\xi << 1$, $T < T_c$, is the macroscopic region in the ordered phase; it refers to phenomena occurring over distances r large compared to ξ. Similarly there is a macroscopic region for $T > T_c$ denoted by III on Fig. 1. The region marked II, in which $k\xi >> 1$ for either $T \geq T_c$ or $T \leq T_c$, is the so-called critical region, which describes phenomena occurring over distances small compared to ξ, but large compared to all other relevant lengths. Since $\xi \to \infty$ as $T \to T_c$, region II is a "macroscopic" region, except insofar as critical fluctuations are concerned.

The static correlation function $C^{\Psi}(\vec{k})$ diverges at the origin $(k = 0, \; \xi^{-1} = 0)$ and remains finite for finite k at $T = T_c$. The scaling hypothesis rests on the assumption that $C^{\Psi}(\vec{k})$ varies smoothly throughout the (k, ξ^{-1}) plane, except for the singularity at the origin. Furthermore, the function is assumed to be essentially determined by its limiting behavior in the three shaded asymptotic regions. Thus, if the forms valid in regions I and II are separately extrapolated to the line L_1 $(k\xi = 1, \; T < T_c)$, then the two resulting expressions must coincide, up to a possible factor of order unity. This means that there is no other dividing line between macroscopic and critical behavior than that provided by the length ξ . Moreover, a single function describes the correlations in the whole (k, ξ^{-1}) plane, with a characteristic dependence on the parameter $\xi/r \sim k\xi$. The susceptibility $\chi_T^{\Psi} = (k_B T)^{-1} C_T^{\Psi}(0)$, defined as a function of temperature along the abscissa of Fig. 1, has a singularity at the point $\xi^{-1} = 0$. For finite k, the correlation function $C_T^{\Psi}(\vec{k})$ passes (along L_3, say) from "below T_c" (region I), through a critical region whose size depends on k, to "above T_c" (region III), with no divergence.

The dynamic scaling hypothesis may be formulated in similar fashion, by focusing on the normal modes which dominate the frequency spectrum of the correlation function $C^{\Psi}(\vec{k},\omega)$. Let $\omega^{\Psi}(\vec{k})$ describe such a normal mode, which can be either "propagating" (Fig. 2a) of "diffusive" (Fig. 2c). In hydrodynamic regions $(k\xi << 1)$, the shape of the frequency spectrum and the form of the function $\omega^{\Psi}(\vec{k})$ may be deduced from an appropriate macroscopic hydrodynamic theory.[17] The dynamic scaling hypothesis asserts that the function $\omega^{\Psi}(\vec{k})$ is fully characterized by its behavior in the three limiting regions of Fig. 1.

In order to give a more precise statement of the scaling

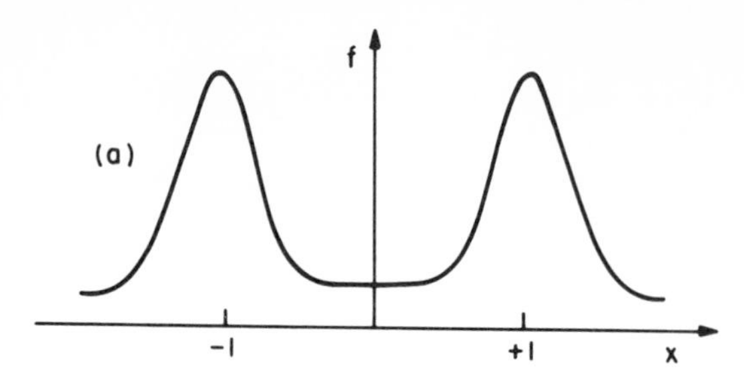

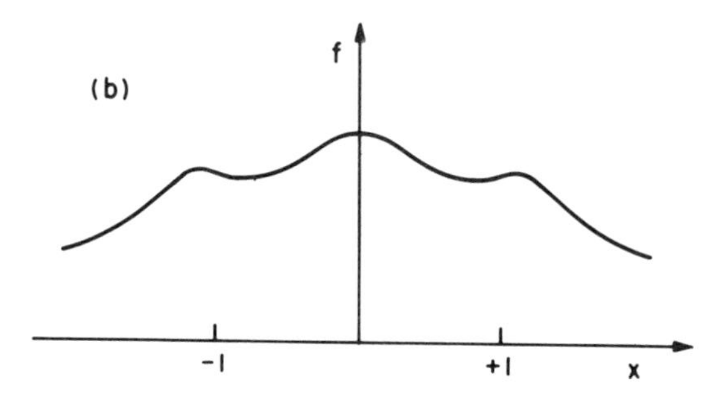

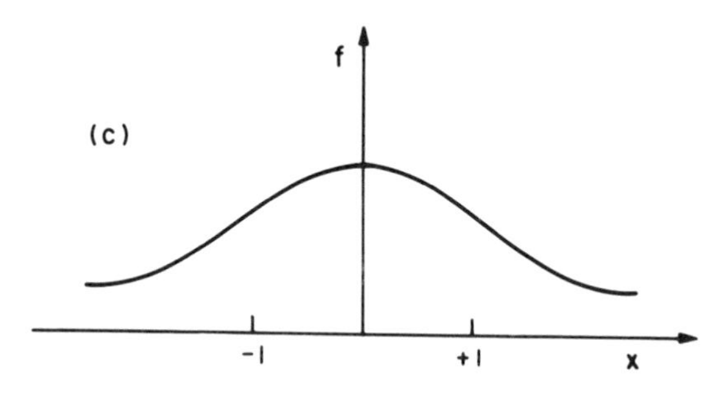

FIG. 2

Some possible forms for the shape function $f^{\vec{N}}(x)$ in the antiferromagnet at T_c (region II), which are all consistent with scaling. Only the shape shown in (a) can be interpreted in terms of propagating spin waves. (Reprinted from Ref. 17.)

hypotheses, we note that a correlation function can always be rewritten in the form

$$C_T^A(\vec{k},\omega) = \frac{2\pi C_T^A(\vec{k})}{\omega_T^A(\vec{k})} \cdot f^A_{\vec{k},T}(\omega/\omega_T^A(\vec{k})) \tag{5}$$

where Eq. (4) implies that

$$\int_{-\infty}^{\infty} f^A_{\vec{k},T}(x)\,dx = 1 \tag{6}$$

and where the <u>characteristic frequency</u> $\omega_T^A(\vec{k})$ is to be determined by the constraint

$$\int_{-1}^{1} f^A_{\vec{k},T}(x)\,dx = \frac{1}{2} . \tag{7}$$

The subscript T on the various functions is meant to indicate their dependence on $T - T_c$. The static scaling hypothesis states that $C_T^A(\vec{k})$ is a homogeneous function of k and ξ^{-1}. The dynamic scaling hypotheses propose that (a) the characteristic frequency $\omega_T^A(\vec{k})$ is also a homogeneous function of k and ξ^{-1}, and (b) the normalized shape function $f^A_{\vec{k},T}(x)$ depends on k and T only through the product $k\xi$ (and the sign of $T - T_c$). These homogeneous functions have the property discussed earlier of being essentially determined by their limiting behavior in the asymptotic regions of Fig. 1. In order to obtain useful physical information, we must first determine the form of $\omega^{\Psi}(\vec{k})$ in one region, and then apply the scaling hypothesis to find its behavior in another region. Let me illustrate how this is done by considering two examples, the antiferromagnet and superfluid helium.

IV. ISOTROPIC ANTIFERROMAGNET

In an "isotropic" antiferromagnet, the Hamiltonian is invariant under simultaneous rotation of all the spins. A system which seems to be well described by such a Hamiltonian is $RbMnF_3$. The order parameter for the antiferromagnet is the staggered magnetization $\vec{N}$, whose correlation function $C_T^{\vec{N}}(\vec{k},\omega)$ is determined from the neutron scattering cross section for momentum transfer $\hbar k$ measured relative to a point on the magnetic reciprocal lattice.

According to the hydrodynamic theory of spin fluctuations in an antiferromagnet[18] the correlation function $C^{\vec{N}}$ is dominated in region I by spin waves with real frequency

$$\omega(\vec{k}) = ck \tag{8}$$

and a damping rate $\frac{1}{2}Dk^2$ which is negligible in the long wavelength limit. The velocity c is given in terms of thermodynamic quantities and may be shown to be proportional to $\xi^{-1/2}$. The shape function $f_{\vec{k},T}(x)$, in the limit $k\xi \to 0$, becomes a pair of δ-functions at $x = \pm 1$. For finite wavelength, the peaks in $f_{\vec{k},T}(x)$ will have nonzero width. The dynamic scaling hypothesis (b) together with the fact that the damping rate varies as k^2, means that the width of the peaks in the normalized shape function $f^{\vec{N}}$ should be proportional to $k\xi$, and the damping constant D must diverge as $\xi^{1/2}$, when $T \to T_c$. At $T = T_c$, we cannot use the hydrodynamic theory, but the dynamic scaling hypothesis (a) predicts that in region II

$$\omega(\vec{k}) \propto k^{3/2} . \tag{9}$$

The dynamic scaling hypothesis does not say whether or not there will be "spin waves" in region II, i.e., whether $f^{\vec{N}}_{k\xi=\infty}(x)$ will have resolvable peaks at $x \neq 0$ (Fig. 2a), or a single maximum at $x = 0$ (Fig. 2c), or some more complicated shape (Fig. 2b). The neutron scattering experiments of Lau et al.[19] in $RbMnF_3$ provide a test of these predic-

predictions. In Fig. 3 the points represent the neutron cross section at T_N, as a function of ω, for different values of momentum. The solid line is a theoretical curve, in which the characteristic frequency has been scaled as $k^{3/2}$ [cf. Eq. (9)], and the area under each curve has been scaled as the static susceptibility, $C^{\vec{N}}_{T_c}(\vec{k}) \sim k^{-2+\eta}$. The agreement between experiment and theory is seen to be excellent.

In region III, the hydrodynamic analysis predicts a finite relaxation rate in the limit $k \to 0$, with corrections proportional to k^2, since the order parameter is not an exact constant of the motion. Together with the dynamic scaling hypotheses, this gives

$$\omega(\vec{k}) \propto \xi^{-3/2}[1+k^2\xi^2] \ . \tag{10}$$

At $k = 0$ the relaxation frequency should be roughly proportional to $T - T_c$, when $T \to T_c^+$. The experimental results[19] for $\omega(\vec{k} = 0) \equiv \Gamma_o$ in $RbMnF_3$ are shown in Fig. 4, where it is seen that Γ_o is indeed proportional to $\xi^{-3/2} \equiv \kappa^{3/2}$, to within the accuracy of the experiment. The behavior of Γ_o may be compared to the Van Hove prediction,[10] which becomes, for the antiferromagnet,

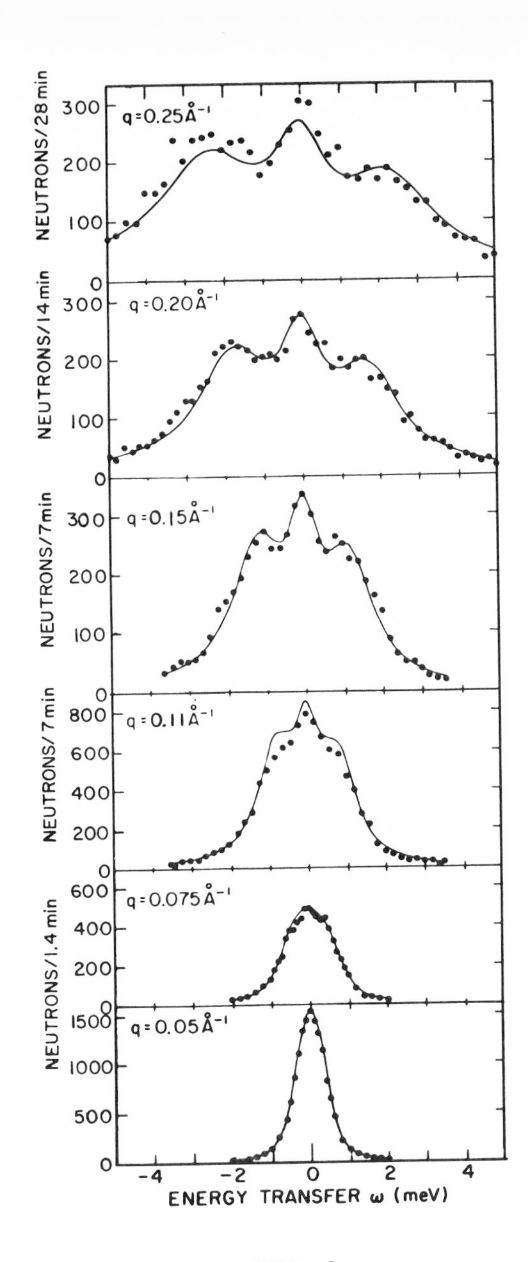

FIG. 3

Neutron scattering cross section for $RbMnF_3$ at T_N, as a function of energy transfer, or frequency, for various momentum transfers q (called k in the text). The points are the data of Ref. 19 and the lines are a fit to the theory of Ref. 17, with parameters adjusted to fit the top curve. (reprinted from Ref. 19.)

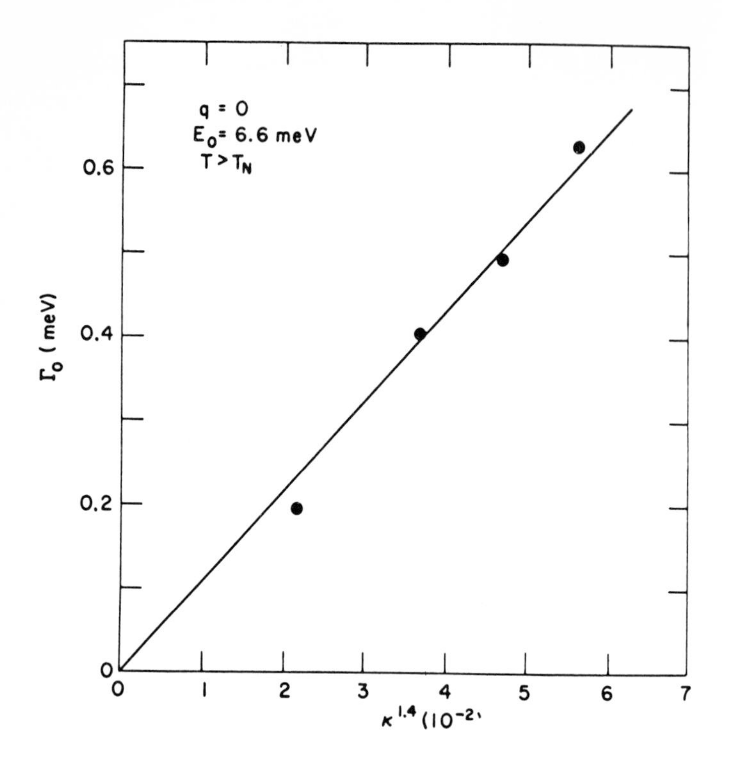

FIG. 4

The characteristic frequency $\omega^{\vec{N}} = \Gamma_o$ at $k = 0$ [Eq. (10)] in $RbMnF_3$, as a function of temperature, measured in units of $\kappa = \xi^{-1}$. (Reprinted from Ref. 19.)

$$\Gamma_o = \Lambda/\chi^s \propto 1/\chi^s \propto \kappa^2 , \qquad (11)$$

where χ^s is the staggered susceptibility, which goes roughly[6,9] as $\xi^2 \propto \kappa^{-2}$. In the scaling theory, the Onsager coefficient Λ, which was taken as a constant by Van Hove, can be seen to diverge as $\kappa^{-1/2}$, yielding the result $\Gamma_o \propto \kappa^{3/2}$.

V. SUPERFLUID HELIUM

The order parameter for the superfluid transition is the Bose field ψ, and is not measurable experimentally.[20] In region I, the theory of two-fluid hydrodynamics[20,21] establishes that $C_T^{\psi}(\vec{k}, \omega)$ is dominated by second sound peaks. The latter have a real frequency $\pm c_2 k$ and a damping rate $\frac{1}{2} Dk^2$, with

$$c_2 \propto \xi^{-1/2} c_p^{-1/2} , \qquad (12)$$

where c_p, the specific heat, has a logarithmic divergence at T_c, and ξ is known to diverge as $(T_c - T)^{-0.67 \pm 0.01}$.[22]

The dynamic scaling hypothesis predicts that the second sound damping coefficient D_2 diverges as $T \rightarrow T_c$:

$$D_2 \propto c_2 \xi , \qquad (13)$$

or, neglecting logarithms, $D_2 \propto |T - T_c|^{-1/3}$. The velocity and damping of second sound can be observed by studying the propagation of heat fluctuations; the correlation function for the "heat density," like that of the order parameter, is dominated by second sound in region I. The experimental results of Tyson[23] for the second sound damping are shown in Fig. 5, along with the theoretical prediction, Eq. (13).

Outside of region I, the correlation functions for the order parameter and the heat density are not simply related, and there is no known way of studying $C^{\psi}_{T}(\vec{k}, \omega)$ directly. However, the dynamic scaling hypothesis as applied to the heat density can be tested in region III, where the hydrodynamic theory of a normal fluid predicts heat diffusion, with

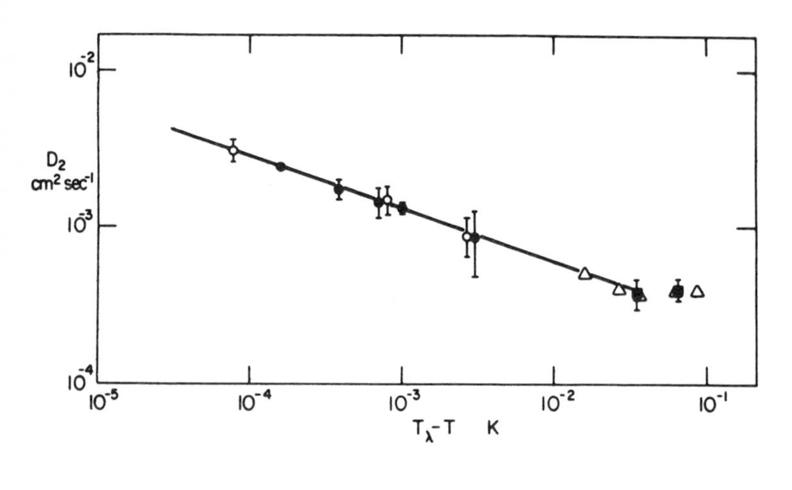

FIG. 5

The damping constant of second sound in liquid He II vs. temperature near T_λ. The circles are the data of Ref. 23, and the solid line has a slope -1/3, as predicted by Eq. (13). (Reprinted from Ref. 23.)

a relaxation rate $D_T k^2$. According to dynamic scaling, we must have

$$D_T \propto c_2 \xi \quad , \tag{14}$$

where D_T and c_2 are evaluated above and below T_c respectively, at the given value of ξ. The heat diffusion constant D_T can be obtained by measuring the thermal conductivity κ, since

$$\kappa = D_T \rho c_p \quad . \tag{15}$$

Neglecting logarithms, κ should diverge as $(T - T_c)^{-1/3}$, as $T \to T_c$ from above. The divergence in κ was first seen qualitatively by Kerrisk and Keller,[24] and has been investigated quantitatively by Ahlers,[25] whose results are shown in Fig. 6. The solid line is the theoretical value of κ, obtained from Eqs. (14) and (15), where is it assumed that the exponent ν of ξ for $T > T_c$ is equal to $2/3$, as measured for $T < T_c$.[17,22] The agreement between experiment and theory is clearly spectacular, and provides one of the most detailed verifications to date of the correctness of the scaling laws. In particular, it should be noted that the scaling of frequencies expressed in Eq. (14) is precise enough to predict the departure of the thermal

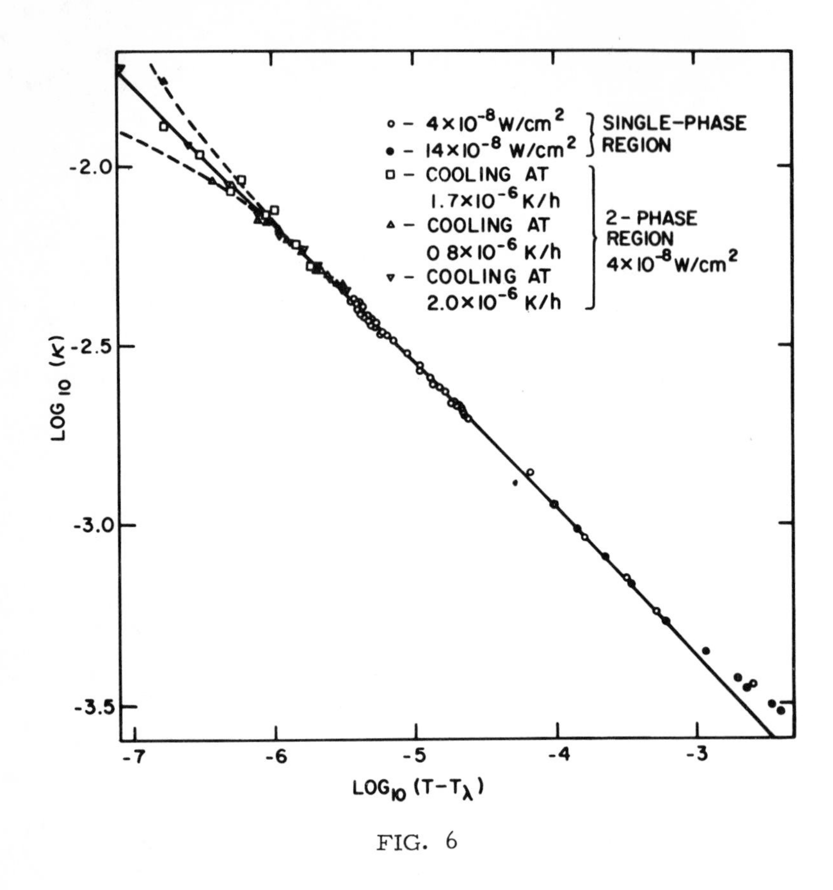

FIG. 6

The thermal conductivity of liquid He I vs. temperature near T_λ. The symbols are the data of Ref. 25 and the solid line is the theoretical prediction of Eqs. (14) and (15), using the measured values for c_p, c_2, and $\xi \propto \rho_s^{-1}$. (Reprinted from Ref. 25.)

---

conductivity from pure power-law behavior, which occurs as a consequence of the terms involving $\log(T - T_c)$. Indeed, if an "apparent" exponent is defined for each decade of reduced temperature, then the presence of the logarithmic factors in Eq. (14) will lead to an exponent which changes with temperature, and only reaches the value 1/3 asymptotically as $T \rightarrow T_c$. A comparison of the apparent exponents from theory [Eq. (14)] and experiment is shown in Fig. 7, from which it is seen that the theory correctly predicts the departures from asymptotic behavior. The average apparent exponent obtained from a power-law fit to the data of Fig. 6 is 0.45, rather than 0.33.

VI. CONCLUSION

I hope that these examples will be sufficient to convince you that the scaling theory is a definite improvement over the Van Hove approach. Moreover, experimental investigations of nonequilibrium phenomena seem to be catching up in precision and in scope with measurements of static properties. In a sense, therefore, I believe that experiment and phenomenology have had their say in the matter, and the time is ripe for more serious theoretical developments, i.e., for "hard theory" and numerical experiments. I do not mean, of course, that experimentalists should stop working, since they may very well embarass us and find systems in which the

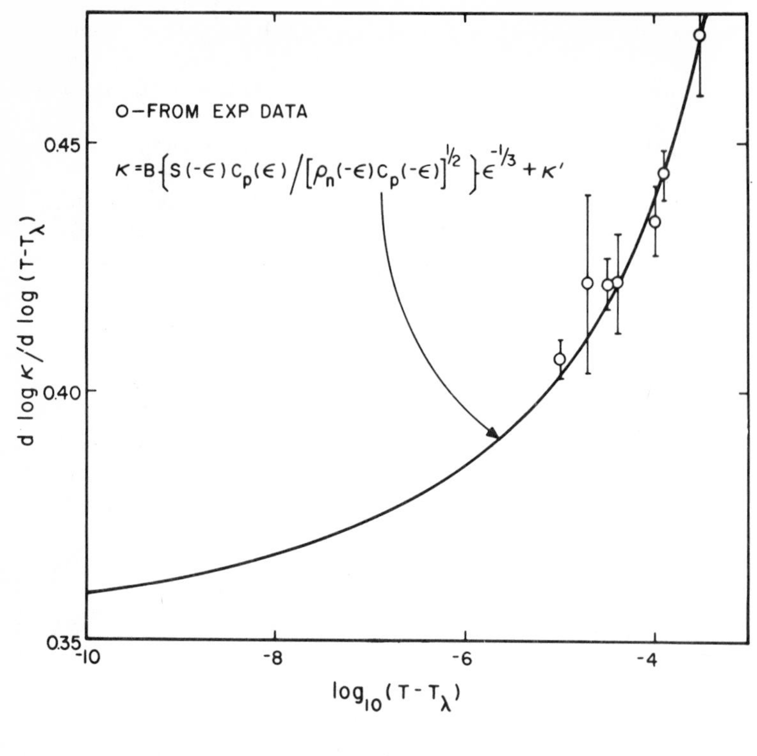

FIG. 7

The apparent critical exponent of the thermal conductivity vs. temperature. The circles indicate exponents obtained from a fit to the data in limited ranges of $T - T_c$, and the solid line is the result obtained from Eqs. (14) and (15). (Reprinted from Ref. 25.)

---

phenomenology is not adequate. But in the systems studied so far, where the theory fits so beautifully, it seems unlikely that the accuracy of experiments can be improved sufficiently to change our understanding in a fundamental way, even if small discrepancies are discovered. On the other hand, I would like to suggest to this distinguished audience of "hard theorists," that the field of dynamic critical phenomena deserves their attention, that it has come of age, and offers mathematical physicists an important challenge in the coming decade.

REFERENCES

1. The bibliography presented here is by no means complete. The reader interested in further information should consult Refs. 6, 7, 9, and 17 below, and L. P. Kadanoff, J. Phys. Soc. Japan, (Suppl.), 26, 122. (1969).

2. L. D. Landau and E. M. Lifshitz, Statistical Physics, Addison-Wesley Publishing Co., Inc., Mass. (1958). Ch. XIV.

3. L. Onsager, Phys. Rev., 65, 117 (1944).

4. W. M. Fairbank, M. J. Buckingham, and C. F. Kellers, Proc. 5th Int. Conf. Low Temp. Phys., Madison, Wisconsin, Milwaukee, Univ. of Wisconsin Press, 1957.

5. M. I. Bagatski, A. V. Voronel, and V. G. Gusak, Zh. Eksperim. i Teor. Fiz., 43, 728 (1962) (English translation), Sov. Phys. JETP, 16, 517 (1963).

6. See M. E. Fisher, Rep. Prog. Phys., 30, 615 (1967).

7. See P. Heller, Rep. Prog. Phys., 30, 731 (1967).

8. B. Widom, J. Chem. Phys., 43, 3892, 3898 (1965).

9. L. P. Kadanoff, et al., Rev. Mod. Phys., 39, 395 (1967).

10. L. Van Hove, Phys. Rev., 95, 1374 (1954).

11. B. Jacrot, J. Konstantinovic, G. Parette, and D. Cribier, Inelastic Scattering of Neutrons in Solids and Liquids, Int. Atomic Energy Agency, Vienna, 1963, p. 317; L. Passell, K. Blinowski, T. Brun, and P. Nielsen, Phys. Rev., 139, A1866 (1965).

12. W. Marshall, Proc. Wash. Conf. (M. S. Green and J. V. Sengers, eds.), N. B. S. Misc. Publ., 273, 135 (1966).

13. M. Fixman, J. Chem. Phys., 47, 2808 (1967), and references therein.

14. K. Kawasaki and M. Tanaka, Proc. Phys. Soc. (London), 90, 791 (1967); K. Kawasaki, Progr. Theoret. Phys. (Kyoto), 40, 9706 (1968), and references therein.

15. L. P. Kadanoff and J. Swift, Phys. Rev., 166, 89 (1968); Ann. Phys. (N. Y.), 50, 321 (1968).

16. R. A. Ferrell, N. Menyhard, H. Schmidt, F. Schwabl, and P. Szepfalusy, Phys. Rev. Letters, 18, 891 (1967); Ann. Phys. (N. Y.), 47, 565 (1968).

17. B. I. Halperin and P. C. Hohenberg, Phys. Rev. Letters, 19, 700 (1967); Phys. Rev., 177, 952 (1969).

18. B. I. Halperin and P. C. Hohenberg, Phys. Rev., 188, 898 (1969).

19. H. Y. Lau, L. M. Corliss, A. Delapalme, J. M. Hastings, R. Nathans, and A. Tucciarone, Phys. Rev. Letters, 23, 1225 (1969).

20. P. C. Hohenberg and P. C. Martin, Ann. Phys. (N. Y.), 34, 291 (1965).

21. I. M. Khalatnikov, Introduction to the Theory of Superfluidity, W. A. Benjamin, Inc., New York, 1965.

22. J. R. Clow and J. D. Reppy, Phys. Rev. Letters, 16, 887 (1966); J. A. Tyson, Phys. Rev., 166, 166 (1968).

23. J. A. Tyson, Phys. Rev. Letters, 21, 1235 (1968).

24. J. F. Kerrisk and W. E. Keller, Phys. Rev., 177, 341 (1969).

25. G. Ahlers, Phys. Rev. Letters, 21, 1159 (1968).

SUBJECT INDEX

W